Grade 3

Reveal MATH®

Differentiation Resource Book

mheducation.com/prek-12

Send all inquiries to:
McGraw Hill
8787 Orion Place
Columbus, OH 43240

ISBN: 978-1-26-421063-3
MHID: 1-26-421063-9

Printed in the United States of America.

8 9 10 11 LHN 27 26 25 24 23

Grade 3

Table of Contents

Unit 2

Use Place Value to Fluently Add and Subtract within 1,000

Lessons

Unit 3

Multiplication and Division

Lessons

Unit 4

Use Patterns to Multiply by 0, 1, 2, 5, and 10

Lessons

Unit 5

Use Properties to Multiply by 3, 4, 6, 7, 8, and 9

Lessons

Unit 6

Connect Area and Multiplicthination

Lessons

Unit 7

Fractions

Lessons

Unit 8

Fraction Equivalence and Comparison

Lessons

Unit 9

Use Multiplication to Divide

Lessons

Unit 10

Use Properties and Strategies to Multiply and Divide

Lessons

Unit 11
Perimeter

Lessons

Unit 12
Measurement and Data

Lessons

Unit 13
Describe and Analyze 2-Dimensional Shapes

Lessons

Lesson **2-1 • Reinforce Understanding**

Represent 4-Digit Numbers

Name ______________________________

Review

You can use place value to represent 4-digit numbers.

thousands	hundreds	tens	ones
2	1	4	4

standard form: 2,144

written form: two thousand, one hundred forty-four

expanded form: 2,000 + 100 + 40 + 4

Fill in the number represented by the base-ten blocks. Write the number in expanded and word form.

1.

thousands	hundreds	tens	ones

____

____

____

2.

thousands	hundreds	tens	ones

____

____

____

Represent 4-Digit Numbers

Name ______________________________

Use the digits to write a number with the greatest possible value. Then write a number with the least possible value. Write each number in standard form, expanded form, and word form.

5 9 8 6

1. Greatest: __________

Expanded form: ______________________________

Word form: ______________________________

____

2. Least: __________

Expanded form: ______________________________

Word form: ______________________________

____

8 1 5 7

3. Greatest: __________

Expanded form: ______________________________

Word form: ______________________________

____

4. Least: __________

Expanded form: ______________________________

Word form: ______________________________

____

Round Multi-Digit Numbers

Name ______________________________

Review

You can use a number line or place-value to round numbers.
Round 487 to the nearest 10 and nearest 100.

Use a number line to round to the nearest 10.

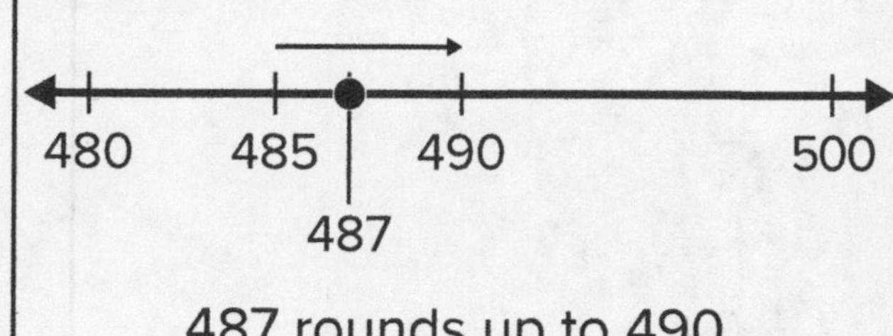

487 rounds up to 490.

Use place value to round to the nearest 100. Look at the tens place when you round to the nearest 100.

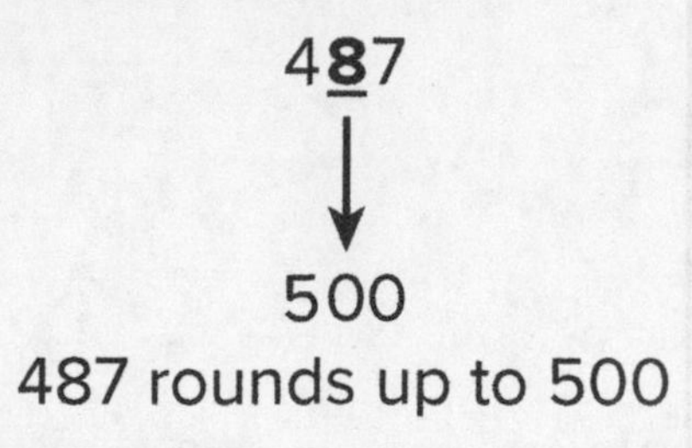

487 rounds up to 500

Use place value to round.

1. Round 468 to the nearest ten. __________

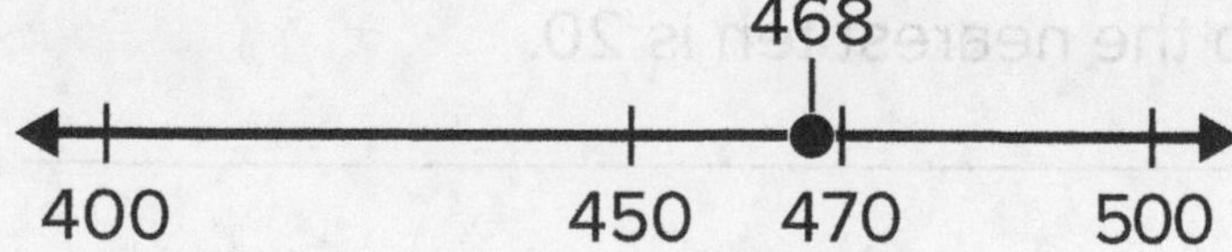

2. Round 468 to the nearest hundred. __________

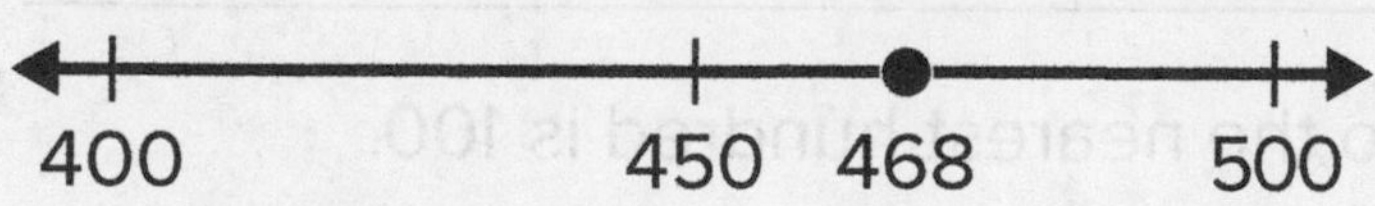

Use place value to round.

3. Round 47 to the nearest ten. __________

4. Round 23 to the nearest ten. __________

5. Round 634 to the nearest hundred. __________

6. Round 219 to the nearest hundred. __________

Round Multi-Digit Numbers

Name ______________________________

Solve.

1. Keisha is shopping. She has $100. She wants to buy colored pencils for $16, 3 sketchpads for $12 each, a set of paints for $15, and a set of paintbrushes for $19. Show how Keisha can use rounding to make sure that she has enough money.

Write three numbers possible for each.

2. A number rounded up to the nearest ten is 20.

____

3. A number rounded down to the nearest ten is 10.

____

4. A number rounded up to the nearest hundred is 100.

____

Estimate Sums and Differences

Name ______________________________

Review

You can use compatible numbers to estimate when an exact sum or difference is not needed.

Mr. Conner spent $122 at the hardware store. Mrs. Basminji spent $276. About how much more did Mrs. Basminji spend?

$276 – $122 = ?

↓ ↓

$275 – $125 = $150

Mrs. Basminji spent about $150 more.

Estimate the difference. Show your work.

1. 298 – 207 = ?

2. ? = 496 – 104

Estimate the sum. Show your work.

3. ? = 416 + 147

4. 274 + 516 = ?

Estimate Sums and Differences

Name ________________________________

Estimate or use compatible numbers to solve. Show your work.

1. Nilda has 536 packages to ship. If she has shipped 94 packages in each of the last 2 hours, about how many packages does she have left to ship?

_____________ packages

Show another way to adjust numbers to solve the problem.

_____________ packages

2. Marcella owns a pack and ship store. Her goal is to move 275 packages a day. If, she has shipped 61 packages in each of the last 3 hours, about how many more packages does she have to move to meet her goal?

_____________ packages

Show another way to adjust numbers to solve the problem.'

_____________ packages

Use Addition Properties to Add

Name ______________________________

Review

You can use properties of addition to help you add.
63 + 30 + 17 = ?

You can switch the order of the addends and the sum will be the same.	You can group the addends in any way and the sum will be the same.
63 + 30 + 17 = 30 + 63 + 17	17 + 63 + 30 = ? 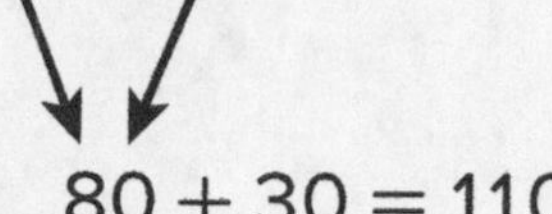 80 + 30 = 110

Complete using properties of addition.

1. 315 + 435 = 435 + ____
2. 273 + 28 + ____ = 56 + 28 + 273
3. ____ + 23 + 12 = 12 + ____ + 72
4. 100 + ____ + 23 = 23 + 407 + ____

Show one way to group the addends to solve. Solve

5. 145 + 13 + 387 = ____

6. 125 + 228 + 72 = ____

Use Addition Properties to Add

Name ______________________________

Solve. Show your work.

1. Mr. Reneke is a manager at the Holiday Hotel and is checking his bank deposit. He is adding $205, $450, and $295. How can he use both properties of addition to add more efficiently?

2. Kiara is checking hotel laundry receipts for one of her customers. His laundry charges for the past three months were $150, $175, and $125. How can she use both properties of addition to add more efficiently?

3. Nestor works at the Holiday House restaurant. He is checking the total cost of the tableware he ordered. He ordered dinner plates for $415, bread plates for $185, and bowls for $160. How can he use both properties of addition to add more efficiently?

Lesson **2-5 • Reinforce Understanding**

Use Addition Properties to Add

Name ______________________________

Review

You can use addition patterns to find and check sums. When you add **even numbers** and **odd numbers,** there are patterns in the sums.

even + even = even	**odd + odd = even**	**even + odd = odd**
14 + 22 = 36	57 + 55 = 112	422 + 177 = 599
342 + 224= 566	247 + 255 = 502	
		odd + even = odd
		383 + 300 = 683

Find and check the sum. Match to the correct statement.

1. 75 + 11 = _____ even = even + even

2. 64 + 14 = _____

3. 29 + 29 = _____ even = odd + odd

4. 42 + 38 = _____

5. 221 + 314 = _____ odd + even = odd

6. 44 + 17 = _____

7. 23 + 14 = _____ even + odd = odd

8. 302 + 101 = _____

Use Addition Properties to Add

Name ______________________________

What Addition Equations Can I Be? My addends are numbers from 24 through 29. My sum is odd. What are the possible equations?

____ + ____ = ____ ____ + ____ = ____

____ + ____ = ____ ____ + ____ = ____

____ + ____ = ____ ____ + ____ = ____

____ + ____ = ____ ____ + ____ = ____

____ + ____ = ____ ____ + ____ = ____

____ + ____ = ____ ____ + ____ = ____

____ + ____ = ____ ____ + ____ = ____

____ + ____ = ____ ____ + ____ = ____

____ + ____ = ____

Use Partial Sums to Add

Name ________________________________

Review

You can use the partial sums strategy to add. One way to break apart numbers is by place value.

addends in a row	addends stacked
158 + 256 = ? 100 + 200 = 300 50 + 50 = 100 8 + 6 = 14 300 + 100 + 14 = 414	1 5 8 + 2 5 6 100 + 200 ⟶ **3 0 0** 50 + 50 ⟶ **1 0 0** 8 + 6 ⟶ + **1 4** **4 1 4**

Break apart each addend. Then add to find the sum.

1. 324 + 135 = ______

______ + 100 = ______

20 + ______ = ______

______ + ______ = ______

______ + ______ + ______ = ______

Stack the addends to find the sum. Show your work.

2. 1 6 3
+ 2 2 5
______
______
+ ______
______

3. 4 5 6
+ 3 2 2
______
______
+ ______
______

Lesson **2-6 • Extend Thinking**

Use Partial Sums to Add

Name ______________________________

Use a 0-9 spinner. Spin to make two 3-digit addends. Show how to use the partial sums strategy to find the sum.

1. ________ + ________ = ________

2. ________ + ________ = ________

3. ________ + ________ = ________

4. ________ + ________ = ________

Decompose to Subtract

Name ____________________

Review

Numbers can be decomposed or "broken apart" in different ways to find the difference.

One Way	**Another Way**
353 − **184** = ?	353 − **184** = ?
353 − **100** = 253	353 − **153** = 200
253 − **80** = 173	200 − **30** = 170
173 − **4** = 169	170 − **1** = 169
353 − 184 = 169	353 − 184 = 169

Pick a number that is easy to subtract from a part when choosing compatible numbers.

Decompose one number to subtract.

1. 302 − 162 = ______

2. ______ = 253 − 132

3. 422 − 233 = ______

4. 383 − 282 = ______

Lesson **2-7 • Extend Thinking**

Decompose to Subtract

Name ______________________________

Roll a number cube to generate two 3-digit numbers. Write a subtraction problem with the greatest number as the first number. Show two ways to decompose one number to subtract.

1. ______ − ______ = ______

2. ______ − ______ = ______

3. ______ − ______ = ______

Adjust Numbers to Add or Subtract

Name ______________________________

Review

Adjust addition and subtraction equation numbers to numbers that are easier to work with.

Subtract from or add the same amount to both numbers in subtraction equations.

Adjust Addition Equations	**Adjust Subtraction Equations**
$316 + 208 = ?$ ↓ −2 ↓ +2 $314 + 210 =$ **524**	$227 - 101 = ?$ ↓ −1 ↓ −1 $226 - 100 =$ **126**

Adjust both numbers to keep the sum or difference the same as the original.

Adjust each equation and find the difference. Show your work.

1. 479 − 98 ______________________________
2. 158 − 46 ______________________________

Adjust each addition equation and find the sum.

3. **367 + 154** ______________________________
4. **543 + 208** ______________________________
5. **224 + 279** ______________________________

Adjust Numbers to Add or Subtract

Name ________________________________

Fans per Team
Jets – 291
Rams – 406
Jaguars – 187
Stars – 388
Bears – 347

A stadium is using game attendance numbers from last year to adjust their seating. The table shows the number of fans per game that attended to support each team.

1. The first game of the year is the Stars versus the Bears. Write an equation showing the number of fans the stadium can expect. Adjust the equation and solve.

____

2. The last game of the year is the Rams versus the Jaguars. Write an equation showing the number of fans the stadium can expect. Adjust the equation and solve.

____

3. For which game should the stadium expect the highest attendance. Why? Write, adjust, and solve an equation to support your answer.

____

4. Which two teams have the most similar attendance? Explain your answer. Write, adjust, and solve an equation to support your reasoning.

____

Use Addition to Subtract

Name ______________________________

Review

You can solve a subtraction problem by writing a related addition equation with an unknown addend. The unknown difference is the same number as the unknown addend.

Solve. 543 − 261 = ?

Think: 261 plus a number equals 543.

Write: 261 + ? = 543

543 − 261 = **282**
261 + **282** = 543

Write a related addition equation for each subtraction equation.

1. 845 − 193 = ? ______________________
2. 679 − 291 = ? ______________________
3. 712 − 436 = ? ______________________
4. 363 − 192 = ? ______________________

Write a related addition equation for each subtraction equation. Solve each problem.

5. 734 − 122 = ? ______________________
6. 591 − 356 = ? ______________________
7. 304 − 277 = ? ______________________
8. 280 − 173 = ? ______________________
9. In a survey, 523 students had 2 siblings and 355 students had 1 sibling. How many more people had 2 siblings? __________

Use Addition to Subtract

Name ______________________________

Pet Ownership
1 pet - 312 people
2 pets - 501 people
3 pets - 474 people
4 pets - 107 people

The table shows the results of a survey about pet ownership.

1. How many more people own 2 pets than 1 pet? Write both an addition and a subtraction equation.

____

2. How many more people own 2 pets than 3 pets? Write both an addition and a subtraction equation.

____

3. Sam uses the survey results to write a third subtraction equation. He uses the related addition equation 312 + ? = 474 to help him find the missing difference. What question is Sam trying to answer? Explain.

____

4. What is a question you could ask about the number of people who own 4 pets? Write both an addition and a subtraction equation to answer your question.

Lesson **2-10 • Reinforce Understanding**

Fluently Add Within 1,000

Name ____________________

Review

You can use different strategies to find a sum.

Partial Sums	Adjust Addends
$\begin{array}{r} 352 \\ +117 \\ \hline \end{array}$ 300 + 100 → 400 50 + 10 → 60 2 + 7 → + 9 **469**	352 + 117 = ? 352 ↓ −2, 117 ↓ +2 350 + 119 = **469**

Use partial sums to find the sum.

1. 258 + 421

Adjust the addends to find the sum.

2. 604 + 242

Use a strategy to find each sum. Show your work.

3. 312 + 524	**4.** 706 + 183
5. 422 + 228	**6.** 293 + 681

Fluently Add Within 1,000

Name ______________________________

On Monday, Kyle's family drove 247 miles. The next day, they drove 184 more miles than the day before. On the final day, they drove 312 miles.

1. How many miles did they drive on Tuesday? Solve the problem using partial sums.

2. How many miles did they drive during the first two days of the trip? Solve the problem using adjusted addends.

3. How many miles did they drive during the entire trip? Use either partial sums or adjusted addends.

Lesson **2-11 • Reinforce Understanding**

Fluently Subtract Within 1,000

Name ________________________________

Review

You can use different strategies to find a sum.

Solve. 459 − 261 = ?

Decompose One Number	Adjust Numbers	Related Addition Equation
459 − 200 259 − 50 209 − 10 199 − 1 198	459 + 1 = 460 261 + 1 = 262 460 − 260 200 − 2 198	459 − 261 = **198** 261 + **198** = 459

Use a strategy to find the difference. Show your work.

1. 253 − 125 = ? ________

3. 456 − 375 = ? ________

2. 867 − 342 = ? ________

4. 598 − 364 = ? ________

Fluently Subtract Within 1,000

Name ________________

Jan sold a total of 752 tickets to baseball games this week. The table shows the number of tickets sold on the days of the week.

Day of the Week	Number of Tickets Sold
Sunday	225
Monday	216
Tuesday	83
Wednesday	47
Thursday	38
Friday	31
Saturday	?

1. Is the difference between the number of tickets Jan sold on Sunday and Wednesday greater than, less than, or equal to the difference between the number of tickets she sold on Monday and Thursday? Explain your answer.

2. How many tickets did Jan sell on Saturday? Show your work.

3. How many more tickets did Jan sell on the day she sold the most tickets than she did on the day she sold the least?

How Can I Solve Two-Step Problems Involving Addition and Subtraction?

Name ________________________________

Review

Julius works at an orchard.for two days. On the first day, he picks 231 apples. On the second day, he picks 16 fewer apples than on first day. How many apples does he pick this week?

Step One	Step Two
How many applesdoes he pick on the second day?	How many apples does he pick this week?
$231 - 16 = a$	$231 + 215 = b$
$231 - 1 = 230$	$231 + 200 = 431$
$16 - 1 = 15$	$431 + 10 = 441$
$230 - 15 = 215$	$441 + 5 = 446$
He picks 215 apples on the second day.	He picks a total of 446 apples this week.

Solve. Show your work.

1. Lola walked Monday, Tuesday, and Wednesday for a total of 270 minutes. She walked 112 minutes on Monday and 86 minutes on Tuesday. How long did she walk Wednesday?

____

2. Liam drove 462 miles Friday. He drove 21 more miles on Saturday than he did on Friday. How many total miles did Liam drive those two days? ________________

How Can I Solve Two-Step Problems Involving Addition and Subtraction?

Name ______________________________

Renee worked Monday through Friday this week for a total of 40 hours. On Monday, she worked six hours. On Tuesday, she worked three more hours than on Monday. On Wednesday, she worked five fewer hours than on Tuesday. On Thursday, she worked seven more hours than on Wednesday.

1. How many hours did Renee work Monday through Thursday? Complete the table.

Day of the Week	Number of Hours Worked
Monday	6
Tuesday	
Wednesday	
Thursday	
Friday	

2. She is paid $8 per hour. How much did she get paid for each day? Complete the table.

Day of the Week	Wages ($)
Monday	48
Tuesday	
Wednesday	
Thursday	
Friday	

3. How much did Renee get paid for the week? ________

Understand Equal Groups

Name ________________________________

Review

When objects are in equal groups, multiplication helps you determine the total.

There are 4 pots with 2 flowers in each pot.

Each pot is one group.

Each flower is one object.

4 equal groups of 2

$4 \times 2 = 8$

Draw equal groups to represent the multiplication.

1. $3 \times 6 = 18$

Write the multiplication equation to represent equal groups.

2.

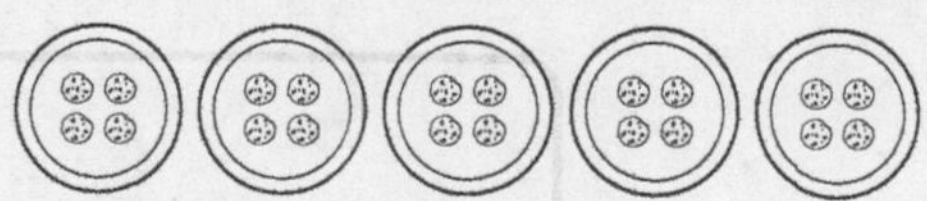

$5 \times$ ____ $=$ ____

3.

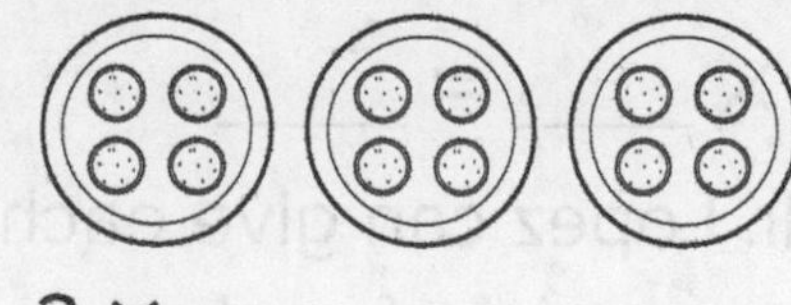

$3 \times$ ____ $=$ ____

Draw equal groups to represent the multiplication. Write the total.

4. $2 \times 4 =$ ____

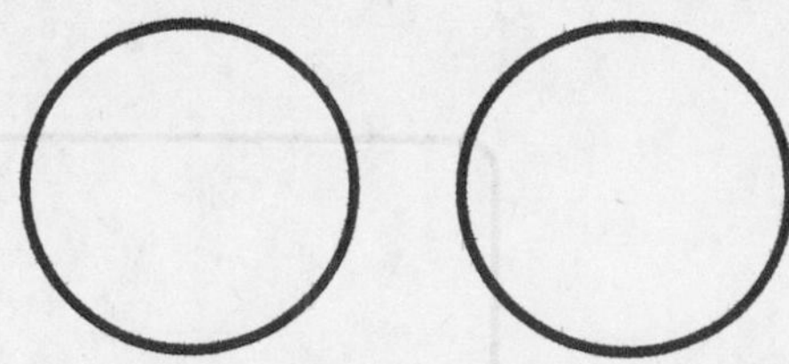

Understand Equal Groups

Name ______________________________

1. Determine different ways to show equal groups of objects with a product of 12. Draw a picture to help. Write a multiplication equation to show each way you find.

2 × ____ = ____

____ × 2 = ____

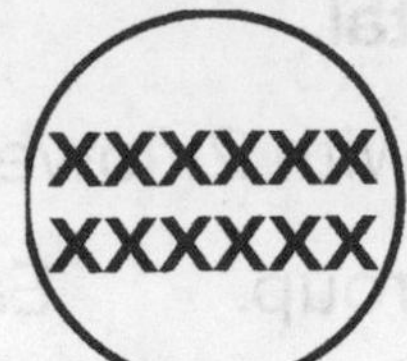

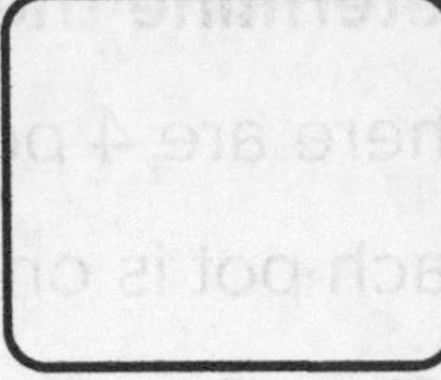

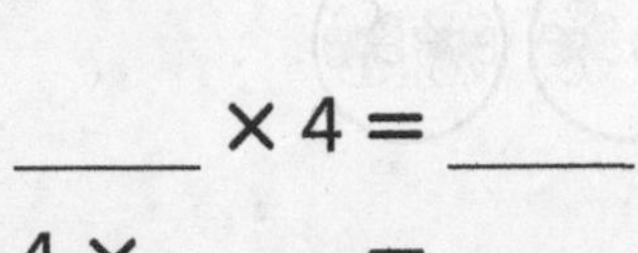

____ × 4 = ____

4 × ____ = ____

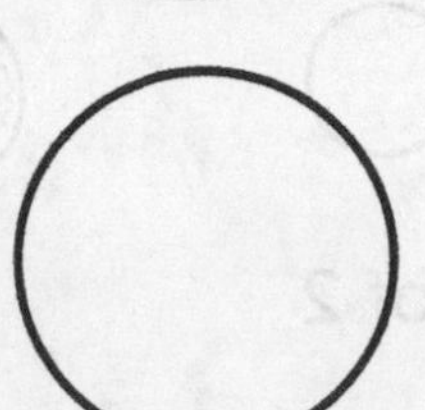

2. Use equal groups to write multiplication equations to solve the problems. Draw a picture to help.

Mr. Lopez is buying socks for 4 grandchildren. There are 12 pairs of socks in a package. He wants to give an equal number of socks to each grandchild. How many pairs of socks can Mr. Lopez give to each grandchild?

4 × ____ = ____

Mr. Lopez can give each grandchild ____ pairs of socks.

How many pairs of socks could Mr. Lopez give to each grandchild gets if the factory added another 4 pairs of socks to the package?

____ + 4 = ____

4 × ____ = ____

Mr. Lopez could give each grandchild ____ pairs of socks.

Lesson **3-2 • Reinforce Understanding**

Use Arrays to Multiply

Name ______________________________

Review

Arrays represent multiplication equations showing equal groups.

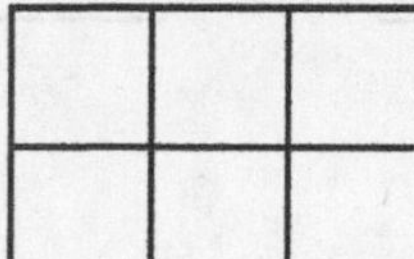

2 equal groups of 3

$2 \times 3 = 6$

factors product

Draw the array that represents the multiplication equation. Circle the factors.

1. $4 \times 3 = 12$

2. $6 \times 3 = 18$

Write the multiplication equation that the array represents and solve.

3.

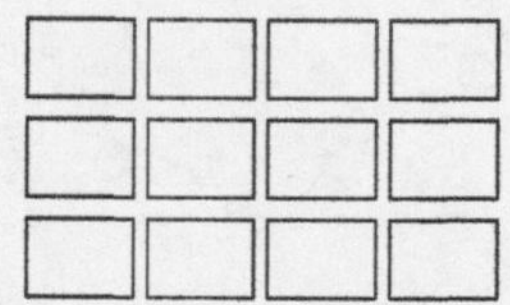

$3 \times$ ____ $=$ ____

4.

____ $\times 5 =$ ____

Lesson **3-2 • Extend Thinking**

Use Arrays to Multiply

Name ______________________________

1. Mr. Grabach is building rectangles with 10 tiles. Draw all the possible arrays that he could make with his tiles. Complete the multiplication equation for each array.

1 × ____ = ____	____ × 1 = ____
____ × 5 = ____	5 × ____ = ____

2. Calvin needs 9 jerseys for each of the 6 teams in his baseball league. How many jerseys does Calvin need to order?

Draw an array. Write the multiplication equation and solve.

9 × ____ = ____

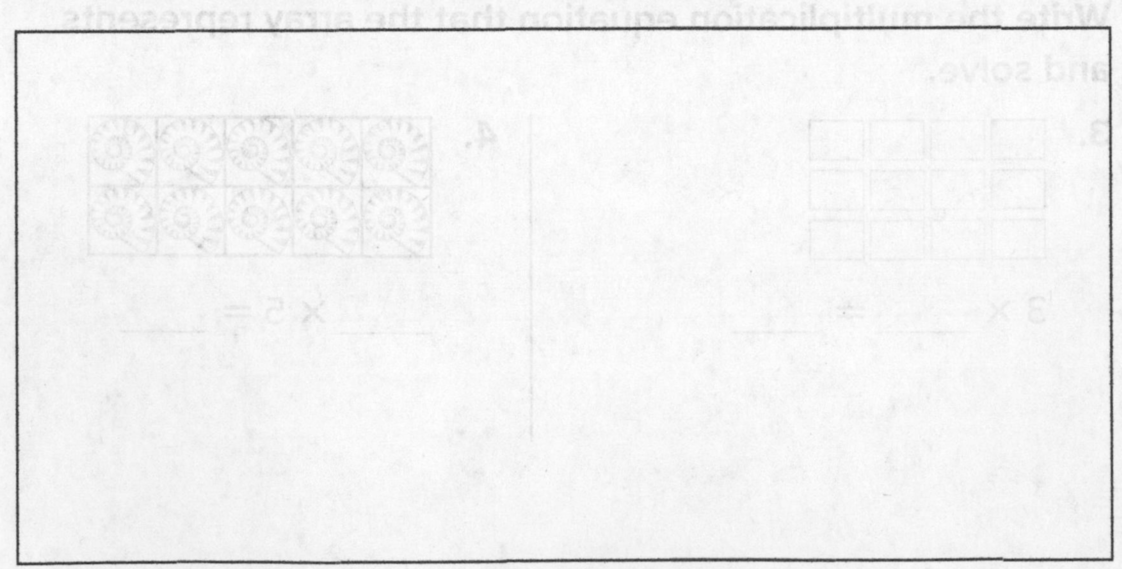

Understand the Commutative Property

Name ____________________

Review

You can multiply two factors in any order and the product stays the same.

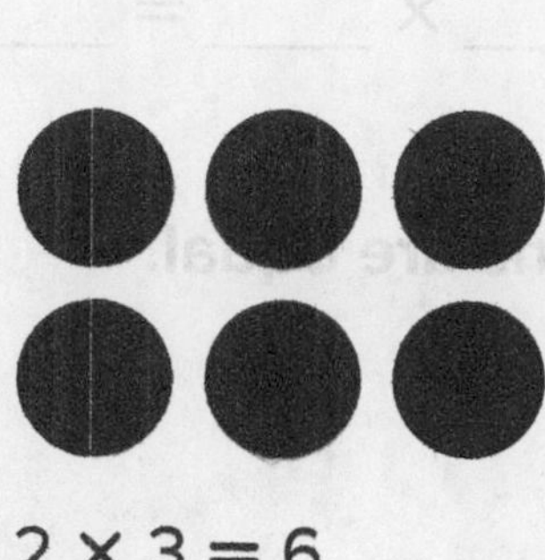

$2 \times 3 = 6$

$3 \times 2 = 6$

Draw arrays to show the expressions are equal.

1. 6×2 and 2×6

2. 3×5 and 5×3

Use a property of multiplication to write Fact 2.

Fact 1	Fact 2
3. $4 \times 2 = 8$	____ × 4 = ____
4. $6 \times 3 = 18$	____ × ____ = 18
5. $7 \times 5 = 35$	5 × ____ = ____
6. $6 \times 9 = 54$	____ × ____ = ____

Lesson **3-3 • Extend Thinking**

Understand the Commutative Property

Name ______________________________

Use a multiplication property to play a game.

Spin the pointer on a spinner twice.
Use the numbers the pointer lands on to write two facts. Multiply.
Write the product.

____ × ____ = ____ ____ × ____ = ____

Draw arrays to show the expressions are equal.

Write all the facts with the product of 24.

1 × ____ = 24	24 × ____ = 24
2 × ____ = 24	____ × 2 = 24
3 × ____ = 24	____ × 3 = 24
____ × 6 = 24	4 × ____ = 24

Understand Equal Sharing

Name ____________________

Review

When you share equally, you divide to make equal groups.

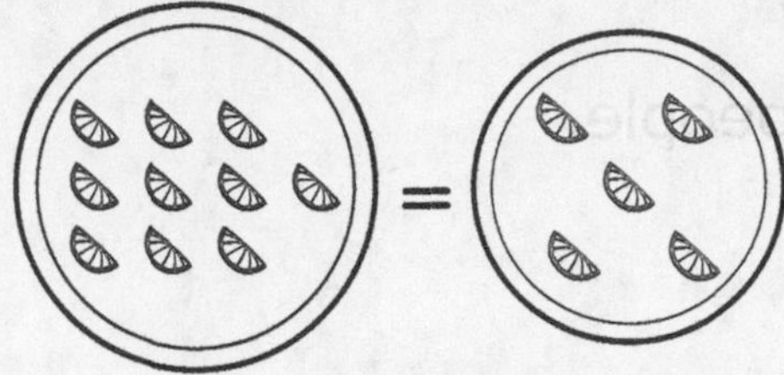

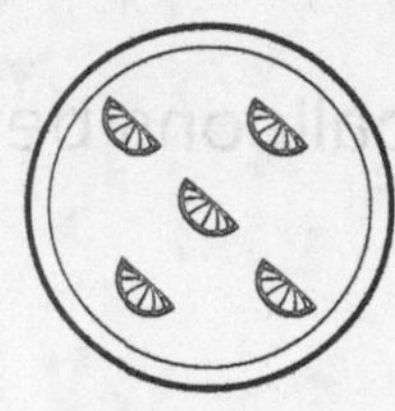

10 slices shared equally into 2 groups equals 5.

$10 \div 2 = 5$

Complete each division equation.

1. $8 \div$ ____ $= 4$

2. ____ $\div 6 = 2$

3. ____ $\div 3 = 3$

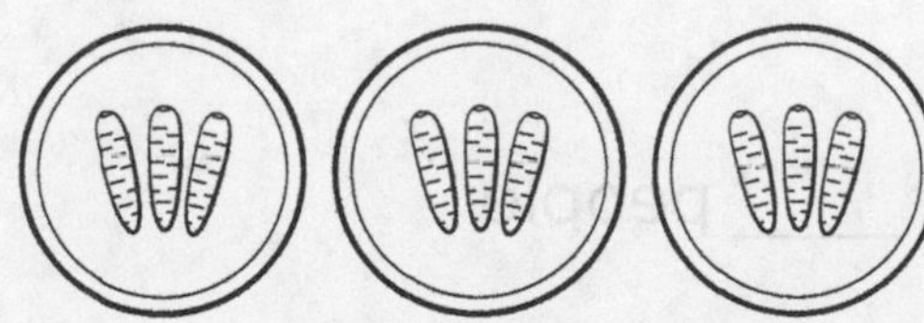

4. ____ $\div 3 = 5$

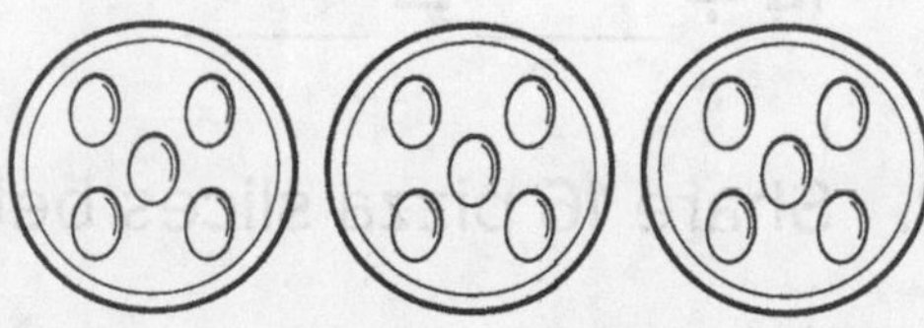

Understand Equal Sharing

Name ____

Your class is planning a party. Find out how many each person will get, if party favors and food are shared equally. Draw a model to show equal shares. Write the division the model represents.

1. Share 12 balloons between _____ people.

12 ÷ _____ = _____

2. Share 14 whistles between _____ people.

14 ÷ _____ = _____

3. Share 16 pizza slices between _____ people

16 ÷ _____ = _____

Understand Equal Grouping

Name ________________________________

Review

Division can represent equal grouping.

These service dogs will visit people in groups of 2. How many groups?

Put the dogs in groups of 2.

3 **equal groups** of 2

$6 \div 3 = 2$

Draw and write the equal groups. Complete the division.

1. 8 flowers

2 flowers for each person

How many people?

_____ **equal groups** of 2

$8 \div 2 =$ _____

2. 10 goldfish

5 goldfish in each pond

How many ponds?

_____ **equal groups** of 5

$10 \div 5 =$ _____

Understand Equal Grouping

Name ____

Find the quotient for each expression. Write a problem for each equation in which equal groups can be expressed.

1. 25 ÷ 5 = _____

2. 63 ÷ 7 = _____

3. 18 ÷ 9 = _____

4. 30 ÷ 6 = _____

Relate Multiplication and Division

Name ______________________________

Review

You can use a bar diagram to model multiplication and division.

3 groups of 2 = 6

6 divided by 3 = 2

Multiplication	Division
$3 \times 2 = 6$ 2 | 2 | 2 6	$6 \div 3 = 2$ 6 2 | 2 | 2

Write a multiplication and division equation for each model.

1.

24		
8	8	8

$24 \div$ _____ = _____

$3 \times$ _____ = _____

2.

7	7	7	7
28			

_____ $\times 7 =$ _____

_____ $\div$ _____ $= 7$

Match each array to the description.

4 groups of 3 = 12

12 divided by 6 = 2

Relate Multiplication and Division

Name ______________________________

Write a word problem that may be represented by the array. Write a multiplication and division equation to represent the problem and solve.

1. 

12 ÷ _____ = _____   _____ × _____ = 12

2. 

7 × _____ = _____   _____ ÷ 7 = _____

3. 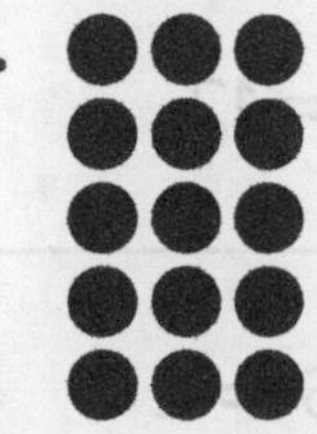

15 ÷ _____ = 3   _____ × 3 = 15

Find the Unknown

Name ______________________________

Review

You can use a number line to model multiplication and division.

$4 \times \boxed{2} = 8$ $8 \div \boxed{4} = 2$

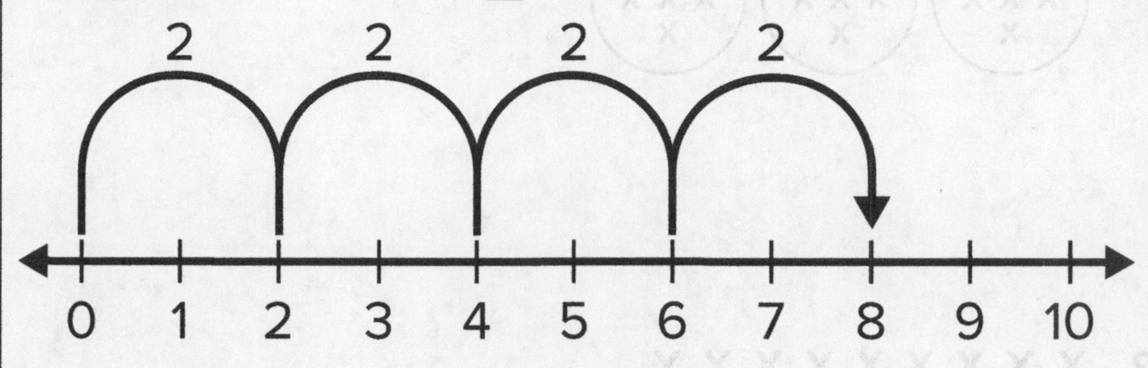

Draw a number line to model each of the following.

1. ? × 3 = 15

2. 18 ÷ 2 = ?

3. 12 ÷ ? = 2

Complete the multiplication and division equation for each model.

4.

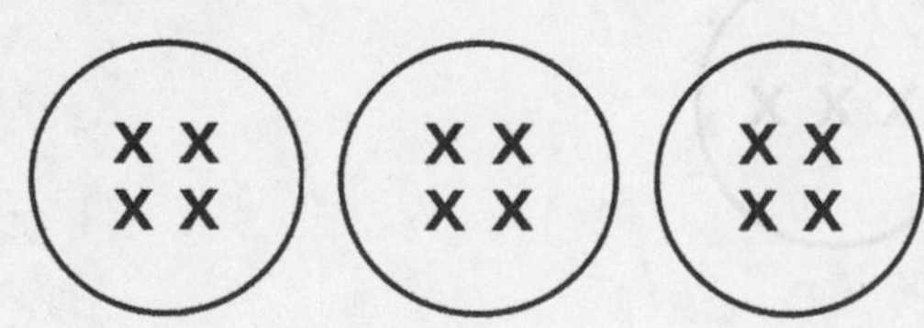

12 ÷ 3 = _____

3 × _____ = 12

5. X X X X X X X X

X X X X X X X X

_____ × 8 = 16

16 ÷ 8 = _____

Find the Unknown

Name ____

Write a word problem that may be represented by the model. Then write a multiplication and division equation to represent the unknown and solve.

1.

X X X X X X X | X X X X X X X | X X X X X X X

2.

3.

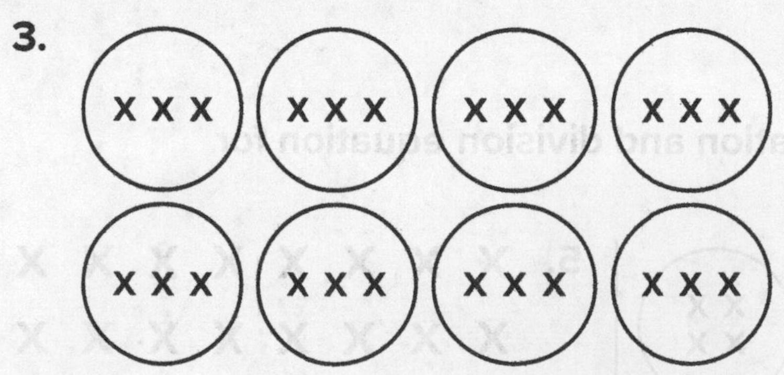

Use Patterns to Multiply with 2

Name ______________________________

Review

You can use patterns to help you recall multiplication facts with 2. Multiplying a number by 2 is the same as doubling the number.

Examples:

$2 \times 4 = \mathbf{8}$ $2 \times 5 = \mathbf{10}$

$4 + 4 = \mathbf{8}$ $5 + 5 = \mathbf{10}$

Complete the equations. Draw lines to show doubling.

1. 2 × ______ = 12	7
2. 2 × ______ = 4	8
3. ______ × 2 = 16	2
4. ______ × 2 = 14	6

Complete the equations.

5. 8 × ______ = 16

6. 2 × ______ = 10

7. 9 × ______ = 18

8. 6 = 2 × ______

9. 2 × ______ = 4

10. 12 = ______ × 2

Use Patterns to Multiply with 2

Name ______________________________

1. Kitty has 2 sisters. Nestor has double that number of sisters. Juana has double Nestor's number of sisters. How can you find the number of sisters Juana has?

2. Irene has 4 grandchildren. David has double that number of grandchildren. Mark has double David's number of grandchildren. How can you find the number of grandchildren Mark has?

3. Dana has 5 cousins. Rami has double that number of cousins. Mohammed has double Rami's number of cousins. Who has more cousins, Mohammed or Rami?

4. Kirit has 3 nephews. Mishaal has double that number of nephews. Asha has double Mishaal's number of nephews. Who has more nephews, Mishaal or Asha?

Use Patterns to Multiply with 5

Name ______________________________

Review

You can use skip counting to help you recall multiplication facts with 5.

HINT: The product of 5 and any number has 0 or 5 in the ones place.

Examples:

5 × 3 = ______ **?**	6 × 5 = ______ **?**
Count: 5, 10, **15**	Count: 5, 10, 15, 20, 25, **30**
5 × 3 = **15**	6 × 5 = **30**

Use skip counting to find the product.

1. 8 × 5 = ______

2. 5 × 7 = ______

3. 6 × 5 = ______

4. 5 × 4 = ______

5. ______ = 5 × 9

6. ______ = 5 × 5

7. ______ = 5 × 1

8. ______ = 5 × 10

Complete the equations.

9. 3 × ______ = 15

10. 5 × ______ = 40

11. 9 × ______ = 45

12. 35 = 5 × ______

13. 1 × ______ = 5

14. 4 × ______ = 20

15. ______ × 5 = 50

16. 2 × ______ = 10

17. 5 × ______ = 30

18. 25 = ______ × 5

Use Patterns to Multiply with 5

Name ______________________________

1. Katrina has 15 cherry tomato seedlings. She wants to plant them in equal rows. Draw arrays to show two different ways Katrina could plant her seedlings in equal rows. Complete the equations.

5 × ________ = 15

________ × 5 = 15

2. Matheus has 5 each of basil, peppers, oregano, and tomato seeds for a pizza garden. He wants to plant the seeds in equal rows. Draw arrays to show two different ways he could plant his seeds. Complete the equations.

5 × ________ = 20

________ × 5 = 20

Use Patterns to Multiply with 10

Name ______________________________

Review

You can use skip counting to help you recall multiplication facts with 10.

Examples:

10 × 5 = **50** 10, 20, 30, 40, **50**

8 × 10 = **80** 10, 20, 30, 40, 50, 60, 70, **80**

Use skip counting to multiply each number by 10.

1. ________ = 10 × 4

2. ________ × 8 = 80

3. 10 × 7 = ________

4. 3 × 10 = ________

Multiply.

5. ________ = 10 × 2 5 × ________ = 10

6. ________ = 10 × 7 5 × ________ = 40

7. 10 × ________ = 30 5 × ________ = 15

8. ________ = 10 × 4 5 × ________ = 20

9. 10 × ________ = 90 5 × ________ = 45

Use Patterns to Multiply with 10

Name ______________________________

1. There are 10 dimes in one dollar. Wendy has 8 dollars in dimes. Maddie has 2 dollars in dimes. How can you find how many dimes Wendy and Maddie have altogether?

2. Sarit has 6 dollars in dimes. Max has 3 dollars in dimes. How can you find how many dimes Sarit and Max have altogether?

3. Pat has 4 dollars in dimes. He has double the number of dimes that Rena has. How can you find the number of dimes Rena has?

4. Wilfren has 5 dollars in dimes. Ashlee has double the number of dimes as Wilfren. How can you find the number of dimes Ashlee has?

Use Patterns to Multiply with 1 and 0

Name ______________________________

Review

You can use patterns to help you recall multiplication facts with 1 and 0.

Think: The **product** of any number multiplied by 1 equals itself. The product of any number multiplied by 0 equals 0.

Examples:

$1 \times 1 = \mathbf{1}$ $1 \times 2 = \mathbf{2}$ $1 \times 3 = \mathbf{3}$

$1 \times 0 = \mathbf{0}$ $2 \times 0 = \mathbf{0}$ $3 \times 0 = \mathbf{0}$

Multiply.

1. $0 \times 4 =$ ________

2. $1 \times 5 =$ ________

3. $0 \times 6 =$ ________

4. $1 \times 7 =$ ________

5. $1 \times 4 =$ ________

6. $0 \times 5 =$ ________

7. $1 \times 6 =$ ________

8. $0 \times 7 =$ ________

Complete the equations.

9. $3 \times$ ________ $= 3$

10. $8 \times$ ________ $= 0$

11. $9 \times$ ________ $= 9$

12. $0 = 5 \times$ ________

13. $8 = 1 \times$ ________

14. $4 \times 1 =$ ________

15. $0 \times 9 =$ ________

16. ________ $\times 10 = 0$

17. ________ $\times 1 = 10$

18. ________ $\times 1 = 6$

Use Patterns to Multiply with 1 and 0

Name ______________________________

Write an equation that follows the pattern of products of 0 or products of 1 to solve each problem. Justify your reasoning.

1. Four friends met to play basketball. Two friends brought 1 basketball each. How many basketballs do they have total?

_______ × _______ = _______

____

____

____

2. Noah goes bowling with 5 friends. If Noah knocks down 0 pins per frame for 10 frames, what will his score be at the end of the game?

_______ × _______ = _______

____

____

____

Multiply Fluently with 0, 1, 2, 5, and 10

Name ____________________

Review

You can use patterns to help you recall facts with 0, 1, 2, 5, and 10.

Think: The product of any number multiplied by 1 equals itself. The product of any number multiplied by 0 equals 0. **Examples:** $1 \times 1 = \mathbf{1}$ $1 \times 0 = \mathbf{0}$	**Think:** Products of facts with a factor of 10 have the same digit in the tens place and a 0 in the ones place. **Examples:** $4 \times 10 = \mathbf{40}$ $5 \times \mathbf{5} = \mathbf{25}$
Think: The product of 2 and any number always has a 0, 2, 4, 6, or 8 in the ones place. **Examples:** $2 \times 1 = \mathbf{2}$ $2 \times 2 = \mathbf{4}$	**Think:** The product of 5 and any number has 0 or 5 in the ones place. **Examples:** $5 \times 1 = \mathbf{5}$ $5 \times 2 = \mathbf{10}$

Complete the equations.

1. $3 \times 1 =$ ______

2. $8 \times$ ______ $= 0$

3. $9 \times 1 =$ ______

4. $0 = 5 \times$ ______

5. ______ $= 10 \times 4$

6. $5 \times$ ______ $= 15$

7. ______ $= 5 \times 4$

8. $9 \times 2 =$ ______

9. ______ $\times 7 = 14$

10. $3 \times$ ______ $= 15$

Multiply Fluently with 0, 1, 2, 5, and 10

Name ______________________________

Use multiplication patterns to answer the questions.

1. Gena drew 3 comic strips. Julia drew double that number. Carina drew double the number that Julia drew. How can you find the number of comic strips that Julia and Carina drew?

____

2. Lana made 24 comic books. She says she can make 5 stacks of her comic books with the same number of comic books in each stack. Is she correct? Explain.

____

3. Mrs. Kim's had 10 boxes of clay. Each box had a 5-pack with different colors of clay. Her students drew comic book characters and copied them on clay to make magnets. They made 2 magnets from each pack of clay using up all 10 boxes. How many magnets did the students make?

____

Solve Problems Involving Equal Groups

Name ______________________________

Review

You can use a multiplication or division to represent and solve problems involving equal groups.

There are 10 kittens in 5 baskets. How many kittens are in each basket?

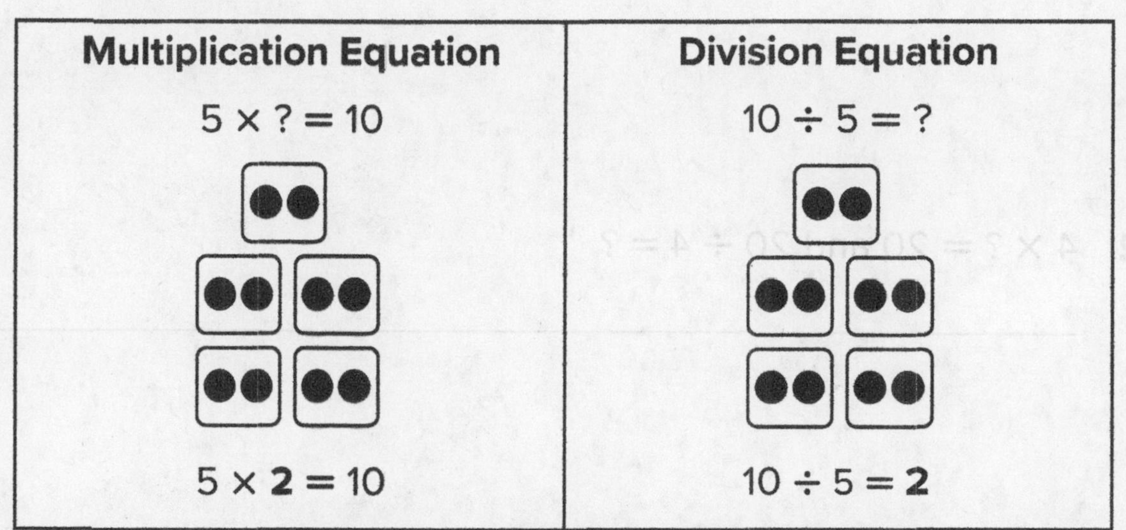

Multiplication Equation	Division Equation
$5 \times ? = 10$	$10 \div 5 = ?$
$5 \times \mathbf{2} = 10$	$10 \div 5 = \mathbf{2}$

Write a multiplication and division equation for each problem.

1. Five orioles share 10 orange slices. If each oriole eats the same amount of orange slices, how many does each bird eat?

   ____

2. A dog walked 20 miles over two days. If he walked the same distance each day, how many miles did he walk each day?

   ____

3. Martina earns $45 per day grooming cats. If she works 5 hours and earns the same amount each hour, how much does she get paid per hour? ______________________________

Solve Problems Involving Equal Groups

Name ______________________________

Describe a situation that could be represented by each set of equations and then solve.

1. $2 \times ? = 18$ and $18 \div 2 = ?$

____

2. $4 \times ? = 20$ and $20 \div 4 = ?$

____

3. $5 \times ? = 50$ and $50 \div 5 = ?$

____

Understand the Distributive Property

Name ______________________________

Review

You can use a property of multiplication to decompose one factor which can make finding the product easier.

Example:

Decompose 4 rows of 3 muffins into two rows of 3 + 3. Then find the products and add them.

$\mathbf{4 \times 3 =} \mathbf{2} \times 3 + \mathbf{2} \times 3$

$\mathbf{4 \times 3 =} \ 6 + 6 = \mathbf{12}$

Solve.

1. ? × 7 = 1 × 7 + 1 × 7 ? = ____

2. ? × 5 = 3 × 5 + 3 × 5 ? = ____

Decompose one of the factors to find the product.

3. 4 × 5 = ____ × ____ + ____ × ____

4 × 5 = ____ + ____ = ____

4. 8 × 4 = ____ × ____ + ____ × ____

8 × 4 = ____ + ____ = ____

Lesson **5-1 • Extend Thinking**

Understand the Distributive Property

Name ______________________________

Use the distributive property.

1. Draw an array. Show how to decompose it into smaller parts. Show how the parts relate to factors and find the product.

2. Explain how decomposing a factor helped you solve 12 × 3 =?

Use Properties to Multiply by 3

Name ________________________________

Review

You can decompose a 3s fact into a 2s fact and a 1s fact. Then add the two products to help you find the product of the 3s fact.

Example: $3 \times 6 = ?$

$2 \times 6 = \mathbf{12}$

$\mathbf{12} + \mathbf{6} = 18$

$1 \times 6 = \mathbf{6}$

Use properties to complete each equation.

1. $3 \times 4 = ?$

$2 \times$ _____ $=$ _____

$1 \times$ _____ $=$ _____

$8 + 4 =$ _____

$3 \times 4 =$ _____

2. $3 \times 5 = ?$

$2 \times$ _____ $=$ _____

$1 \times$ _____ $=$ _____

$10 + 5 =$ _____

$3 \times 5 =$ _____

3. Which expression is equivalent to $21 = 3 \times 7$?

a. $2 \times 7 + 7 \times 7$

b. $2 \times 1 + 7 \times 7$

c. $3 \times 7 + 1 \times 7$

d. $2 \times 7 + 1 \times 7$

4. Which expression is equivalent to $30 = 10 \times 3$?

a. $2 \times 10 + 1 \times 10$

b. $2 \times 1 + 10 \times 10$

c. $3 \times 10 + 1 \times 10$

d. $3 \times 10 + 1 \times 3$

Use Properties to Multiply by 3

Name ______________________________

Show which parts of Carin's work are correct and which are not correct. Explain your answers.

Carin used a property to solve $8 \times 3 = ?$

Carin's Work

$8 \times 3 = ?$

$8 = 4 + 4$

$8 \times 3 = 3 \times 4 + 4$

$= 12 + 4$

$= 16$

Use Properties to Multiply by 4

Name ________________________________

Review

You can decompose a 4s fact into two 2s facts to multiply. Double the product of 2 × 4 to find the product of the the 4s fact.

Example: 4 × 4 = ?

2 × 4 = **8**

8 + **8** = 16

2 × 4 = **8**

4 × 4 = 16

Use properties to complete each equation.

1. 6 × 4 = ?

6 × 2 = _____

6 × _____ = 12

12 + _____ = 24

6 × 4 = _____

2. 4 × 8 = ?

2 × 8 = _____

2 × _____ = 16

16 + _____ = 32

4 × 8 = _____

3. 28 = 4 × _____

2 × 7 = _____

2 × _____ = 14

14 + _____ = 28

28 = 4 × _____

4. _____ = 5 × 4

2 × _____ = 10

2 × 5 = _____

10 + _____ = 20

20 = 5 × _____

Lesson **5-3 • Extend Thinking**

Use Properties to Multiply by 4

Name ____________________

Spin the pointer on a 0-12 spinner. Write a 4s fact using the number the pointer lands on as one factor and 4 as the other factor. Use words and numbers to explain how doubling helps you find the product. Repeat.

Use Properties to Multiply by 6

Name ______________________________

Review

You can decompose a 6s fact to make multiplying easier.

Example 1

You can decompose the factor 6 into 3 and 3.

6 × 5 = ?

★★★★★
★★★★★
★★★★★

3 × 5 = 15

★★★★★
★★★★★
★★★★★

3 × 5 = 15

↓

15 + 15 = 30

6 × 5 = 30

Example 2

You can decompose the factor 6 into 5 and 1.

1 × 5 = 5

6 × 5 = ?

★★★★★
★★★★★
★★★★★
★★★★★
★★★★★

5 × 5 = 25

★★★★★

1 × 5 = 5

25 + 5 = 30

6 × 5 = 30

Complete each equation.

1. 6 × 6 = ?

3 × _____ = _____

3 × _____ = _____

18 + 18 = _____

6 × 6 = _____

2. 4 × 8 = ?

4 × _____ = _____

4 × _____ = _____

16 + 16 = _____

4 × 8 = _____

3. 18 = 6 × ?

5 × 3 = _____

1 × 3 = _____

15 + _____ = _____

18 = 6 × _____

4. ? = 10 × 6

10 × 5 = _____

10 × _____ = _____

_____ + _____ = _____

_____ = 10 × 6

Use Properties to Multiply by 6

Name ____

Roll a number cube. Use the number you roll as one factor and 6 as the other to write a multiplication fact. Explain at least one way to get the answer. Repeat.

Use Properties to Multiply by 8

Name ______________________________

Review

You can decompose an 8s fact in different ways to multiply.

Example 1

Double the product of a 4s fact.

8 × 7 = ?

4 × 7 = 28

4 × 7 = 28

28 × 2 = 5

8 × 7 = 56

Example 2

Think of a 5s fact to help.

8 × 7 = ?

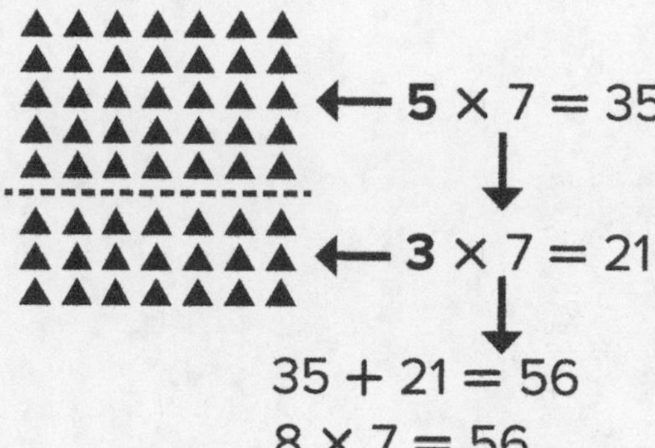

5 × 7 = 35

3 × 7 = 21

35 + 21 = 56

8 × 7 = 56

Complete each equation.

1. 8 × 9 = ?

7 × ____ = ____

1 × ____ = ____

____ + 9 = ____

8 × 9 = ____

2. 6 × 8 = ?

6 × 4 = ____

6 × ____ = 24

____ + 24 = ____

6 × 8 = ____

3. 48 = 8 × ____

4 × ____ = ____

4 × ____ = ____

____ + 24 = ____

8 × 6 = ____

4. ____ = 8 × 8

4 × ____ = ____

4 × ____ = ____

____ + ____ = ____

____ = 8 × 8

Use Properties to Multiply by 8

Name ____

Draw an array model to represent an 8s fact. Draw a line to slice it into two parts. Write the multiplication equations that the parts represent. Show how all equations relate to the model. Repeat with a second array.

Use Properties to Multiply by 7 and 9

Name ______________________________

Review

You can decompose 7s and 9s facts into 5s facts.

Example 1

$7 \times 6 = ?$

Break 7 apart into 5 and 2.

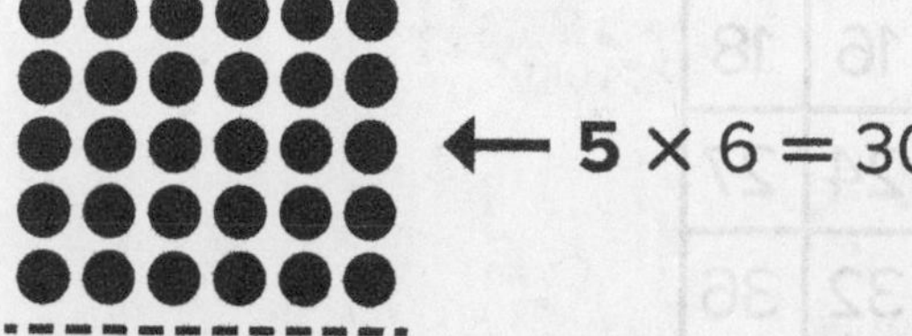

← **5** × 6 = 30

← **2** × 6 = 12

$30 + 12 = 42$

$7 \times 6 = 42$

Example 2

$9 \times 6 = ?$

Break apart 9 into 4 and 5.

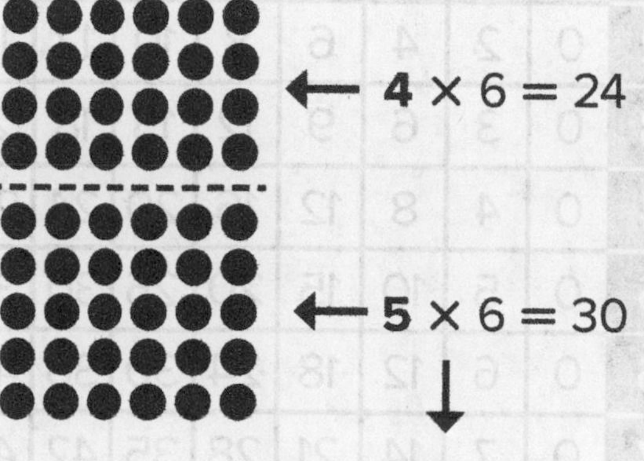

← **4** × 6 = 24

← **5** × 6 = 30

$24 + 30 = 54$

$9 \times 6 = 54$

Complete each equation.

1. $7 \times 8 = ?$

5 × _____ = _____

2 × _____ = _____

_____ + _____ = _____

7 × 8 = _____

2. $9 \times 8 = ?$

4 × _____ = _____

5 × _____ = _____

_____ + _____ = _____

9 × 8 = _____

3. $63 = 7 \times ?$

5 × _____ = 45

2 × _____ = _____

45 + _____ = _____

63 = 7 × _____

4. _____ = 9 × 9

4 × _____ = _____

5 × _____ = _____

_____ + _____ = _____

_____ = 9 × 9

Use Properties to Multiply by 7 and 9

Name ______________________________

Study the multiplication fact table. Look at the rows and the columns. Describe three patterns you find in the 9s facts.

×	0	1	2	3	4	5	6	7	8	9
0	0	0	0	0	0	0	0	0	0	0
1	0	1	2	3	4	5	6	7	8	9
2	0	2	4	6	8	10	12	14	16	18
3	0	3	6	9	12	15	18	21	24	27
4	0	4	8	12	16	20	24	28	32	36
5	0	5	10	15	20	25	30	35	40	45
6	0	6	12	18	24	30	36	42	48	54
7	0	7	14	21	28	35	42	49	56	63
8	0	8	16	24	32	40	48	56	64	72
9	0	9	18	27	36	45	54	63	72	81

What other patterns do you notice in the table?

Solve Problems Using Arrays

Name ________________________________

Review

You can represent a problem by drawing an array and writing an equation. You can break apart one factor to help you solve it.

Problem: A case of juice boxes has 7 rows with 8 boxes in each row. How many boxes of juice are in the case?

Draw an array.

Write an equation. $8 \times 7 = ?$

Break apart one factor.

$8 = \mathbf{4} + \mathbf{4}$

Multiply.

$\mathbf{4} \times 7 = 28$

Double the product.

$\mathbf{28} + \mathbf{28} = 56$ $8 \times 7 = 56$

There are 56 juice boxes in the case.

Solve each word problem. Show your work using an array.

1. An egg carton has 3 rows with 6 eggs in each row. How many eggs are in the carton?

2. A watercolor pan set has 4 rows with 9 pans in each row. How many colors of paint are in the set?

Solve Problems Using Arrays

Name ________________________________

Use the arrays and decomposed factors to help you complete and solve the word problems.

1. Nico planted ____ rows of kale. Each row had ____ kale plants in it. How many kale plants did Nico plant?

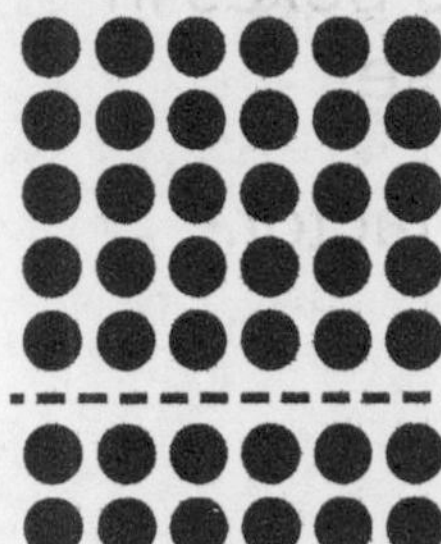

5 × 6 = ____

2 × 6 = ____

____ + ____ = ____

____ × ____ = ____

Nico planted ____ kale plants.

2. Yoshi planted ____ wide rows of pepper plants. Each row had ____ pepper plants. How many pepper plants did Yoshi plant?

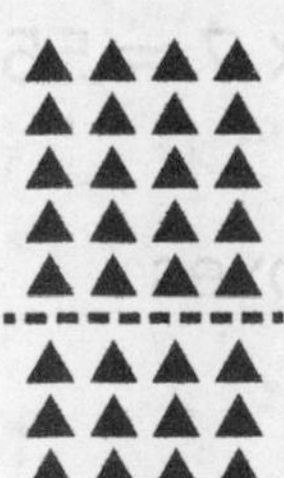

5 × 4 = ____

3 × ____ = ____

____ + ____ = ____

____ × ____ = ____

Yoshi planted ____ pepper plants.

3. Mrs. Martinez planted ____ rows of okra. She planted ____ okra plants in each row. How many okra plants did she plant?

3 × ____ = ____

3 × 5 = ____

____ + ____ = ____

____ × ____ = ____

Mrs. Martinez planted ____ okra plants.

Understand Area

Name ______________________________

Review

Count the square units to find the area of the shaded figure.

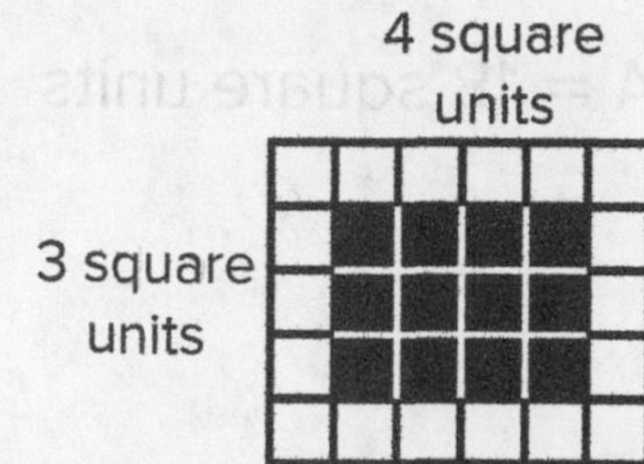

There are no gaps or overlaps. So, there are 12 square units.

The area of the shaded figure is 12 square units.

Find the area of the shaded figures.

1. $A =$ ______ square units

2. $A =$ ______ square units

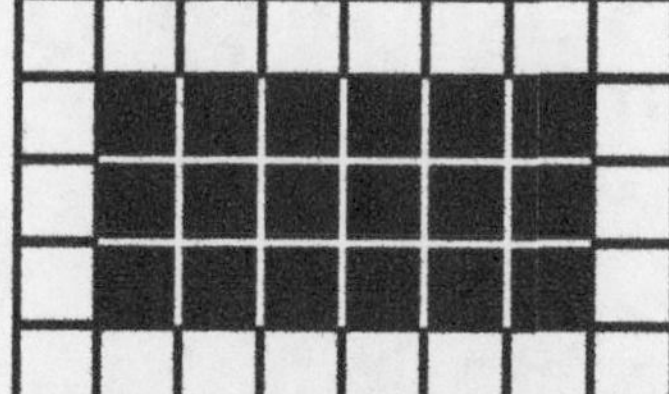

Complete the tiling to find the area of each figure.

3. $A =$ ______ square units

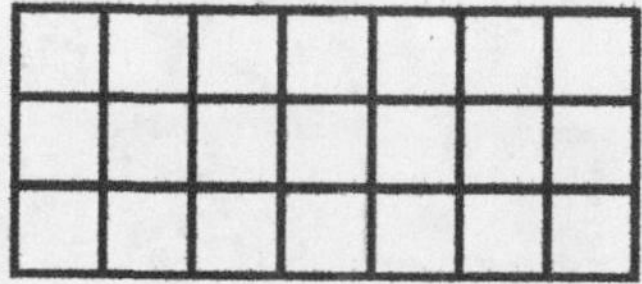

4. $A =$ ______ square units

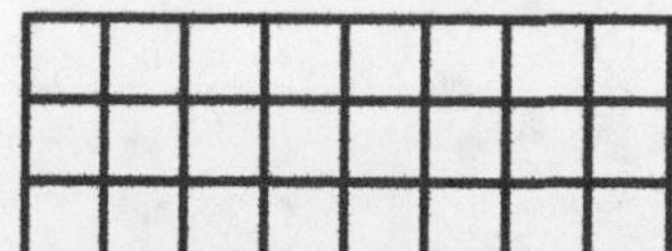

5. $A =$ ______ square units

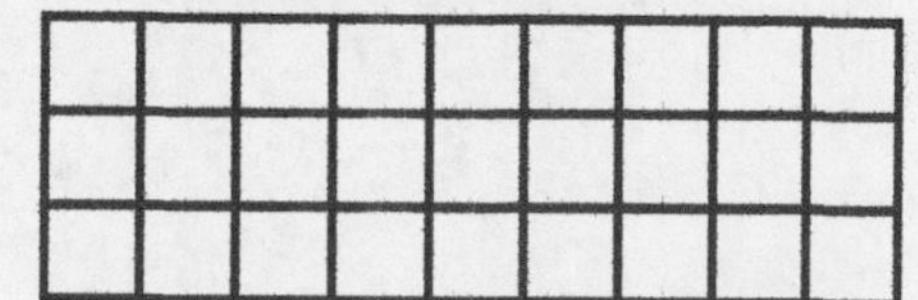

6. $A =$ ______ square units

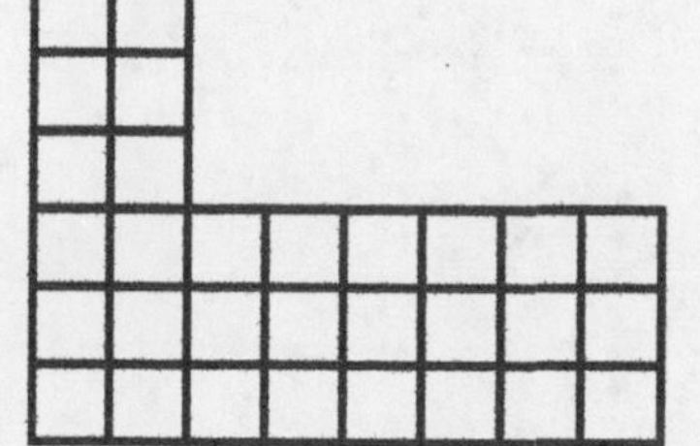

Understand Area

Name ______________________________

Draw three or more different figures with areas of 18 square units. Label the sides of each figure.

$A = 18$ square units

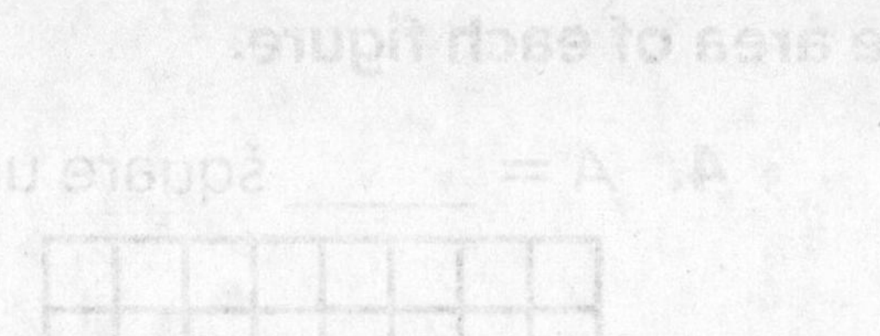

Count Unit Squares to Determine Area

Name ______________________________

Review

You can count square units to find the area of a figure.

2 feet | 3 feet | 1 foot | 1 foot

Remember that area (A) is measured in square units with the same width and length.

6 square units cover the figure. $A = 6$ square feet

Count units to find the area of each figure. Fill in the label when needed.

1. $A =$ _____ square inches

2 in. | 4 in. | 1 in. | 1 in.

2. $A =$ _____ square cm

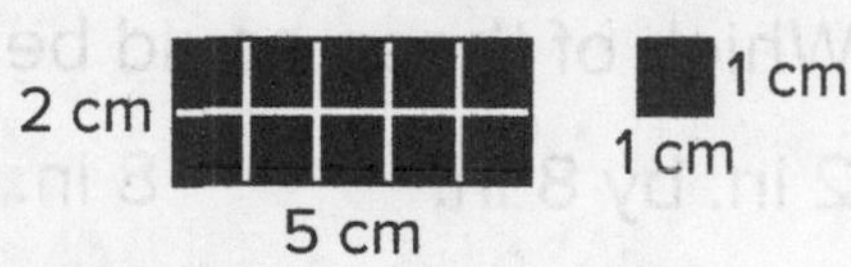

3. $A =$ _____ square _______

4. $A =$ _____ square _______

5. $A =$ _____ square units

6. $A =$ _____ square units

Count Unit Squares to Determine Area

Name ____

Solve each problem. Show your work and explain your answers.

1. Keira has a world map in her room that is 12 square feet. Which of these could be the side lengths of the map?

3 feet by 4 feet   2 feet by 5 feet   2 feet by 6 feet

2. Karl made a map of a park that is 16 square inches. Which of these could be the side lengths of the map?

2 in. by 8 in.   8 in. by 4 in.   4 in. by 4 in.

3. A map of city bus routes folds up so that it is 72 square cm. Which of these could be the side lengths of the folded map?

12 cm by 6 cm   9 cm by 6 cm   9 cm by 8 cm

Use Multiplication to Determine Area

Name ______________________________

Review

You can use multiplication to find the area of the figures.

There are **3 rows** and **5 square units** in each row.

Write an equation to represent the area.

$3 \times 5 = 15$

area = 15 square units

The **length** and the **width** are both **2 inches**.

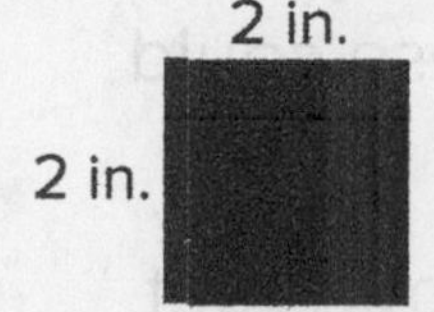

Write an equation to represent the area.

$2 \times 2 = 4$

area = 4 square in.

Multiply to find the area of each figure.

1. $4 \times$ ______ = ______

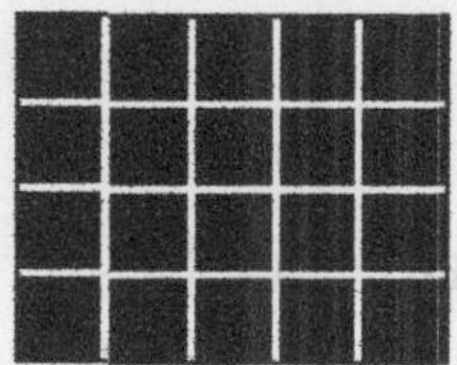

area = ______ square units

2. ______ $\times 8 =$ ______

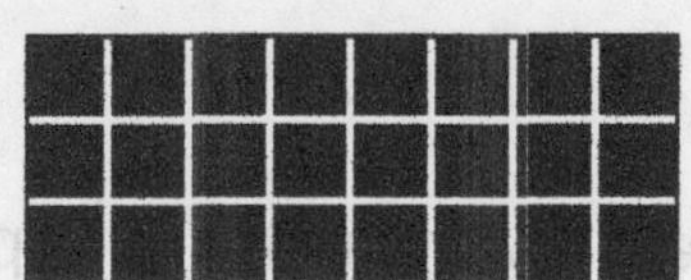

area = ______ square units

3. ______ $\times 4 =$ ______

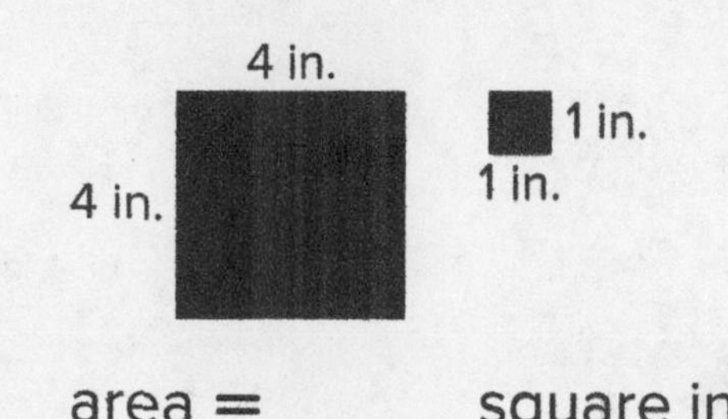

area = ______ square in.

4. $6 \times$ ______ = ______

9 cm

6 cm

area = ______ square cm

Use Multiplication to Determine Area

Name ____

Solve each problem. Show your work by writing an equation.

1. Tristan's garage is the shape of a rectangle. The area of his garage is 24 square feet. Which of these could be the side lengths of the garage?

   8 feet by 3 feet   2 feet by 12 feet   8 feet by 4 feet

2. Connor's basement is in the shape of a square. The area of the basement is 36 square feet. Which of these could be the side lengths of the basement?

   6 feet by 5 feet   6 feet by 6 feet   4 feet by 4 feet

3. Ellen's yard is the shape of a rectangle. The area of her yard is 28 square feet. Which of these could be the side lengths of her yard?

   7 feet by 4 feet   2 feet by 14 feet   7 feet by 3 feet

Determine the Area of a Composite Figure

Name ____

Review

You can decompose a figure into two or more parts. Add the area of the parts to find the total area.

Example 1

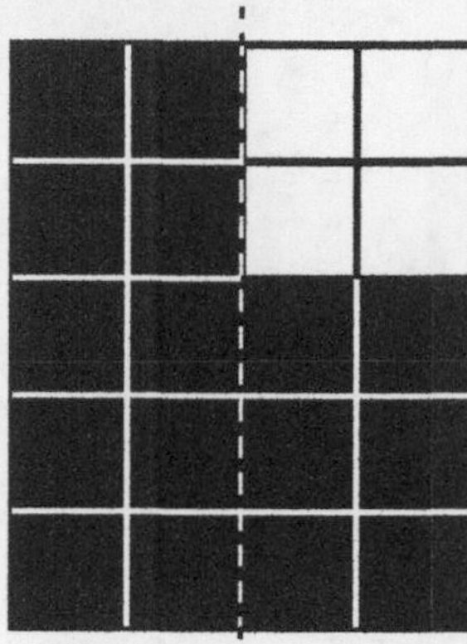

square A = 10 square units
rectangle A = 6 square units

A = 10 + 6 = 16
= 16 square units

Example 2

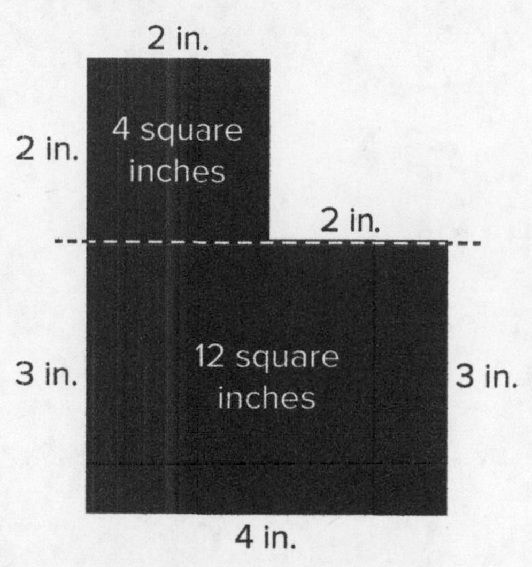

A = 4 + 12 = 16
= 16 square inches

Draw one or more lines to partition each figure. Then find the area of the composite figure.

1.

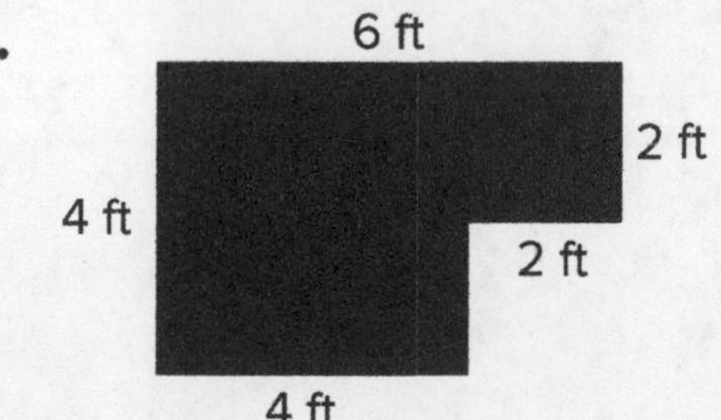

A = _____ + _____ = _____
= _____ square feet

2.

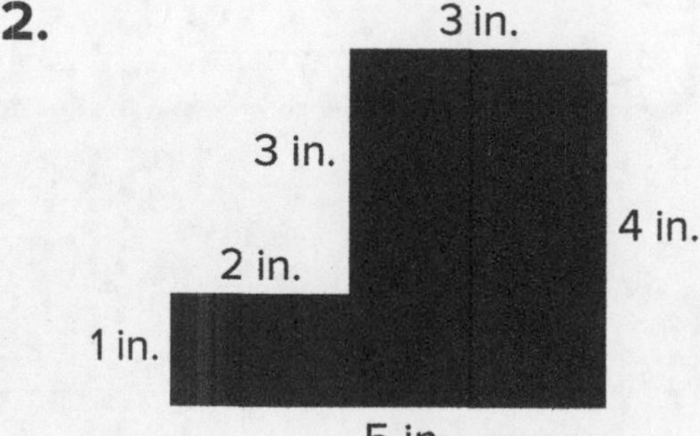

A = _____ + _____ = _____
= _____ square inches

Lesson **6-4 • Extend Thinking**

Determine the Area of a Composite Figure

Name ______________________________

Draw a composite shape with two or more rectangles. Find the total area. Explain what you did as if teaching a new student your way of finding the total area of a composite figure.

Use the Distributive Property to Determine Area

Name ______________________________

Review

You can decompose a side length and add the areas of the parts to find the total area of a rectangle.

Decompose the side length. Find the area of each part.

Add the areas of the parts to find the total area (A).

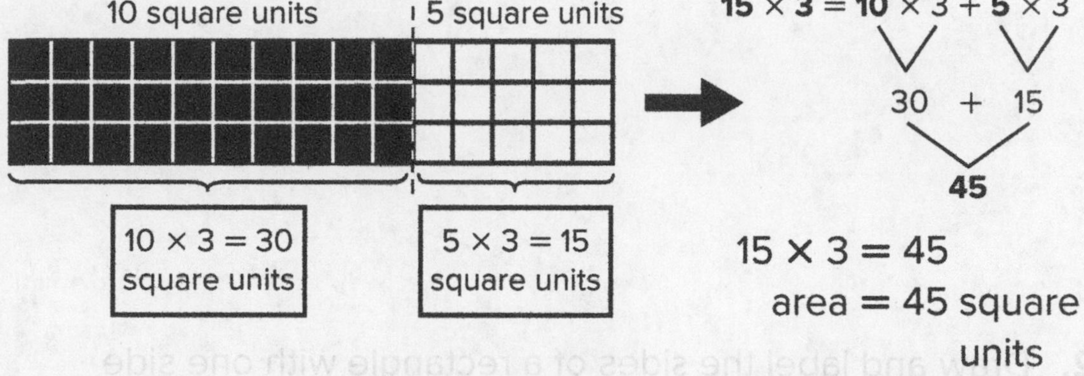

$15 \times 3 = 45$

area = 45 square units

Find the total area of each rectangle.

1. 10 square units | 3 square units | 2 square units

$13 \times 2 = 10 \times$ ____ $+ 3 \times$ ____

= ____ + ____

= ____ square units

2. 7 in. | 2 in. | 4 in.

$7 \times 2 = 7 \times$ ____

$+ 2 \times$ ____

= ____ + ____

= ____ square in.

Use the Distributive Property to Determine Area

Name ____________________

1. Calvin drew a rectangle with one side greater than 10 but less than 20. He multiplies to find its area.

   Calvin's work: $\mathbf{8 \times 3 + 8 \times 10 = 104}$

   Find the dimensions of Calvin's rectangle.
   Explain how you know that your answer is correct.

2. Draw and label the sides of a rectangle with one side greater than 10 in. but less than 20 in. Then complete the equation to find the area.

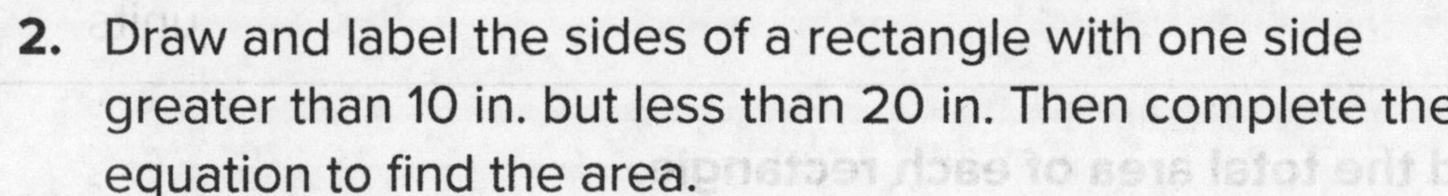

   ____ × ____ + ____ × ____ = ____ square inches

   Explain how you solved the problem.

Solve Area Problems

Name ________________________________

Review

You can use strategies to help you solve area problems. One strategy is to decompose the figure.

Decompose the figure. Multiply the length and width of each part.

$4 \times 1 = 4$
$2 \times 1 = 2$
$1 \times 1 = 1$

Add the 3 areas to find the total area of the figure.

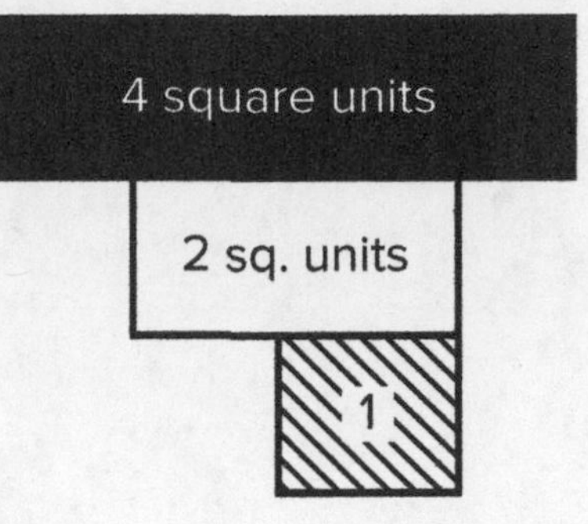

$4 + 2 + 1 = 7$
The area is 7 square units.

Solve the problem.

1. Vicky designs a footpath for her garden. What is its total area?

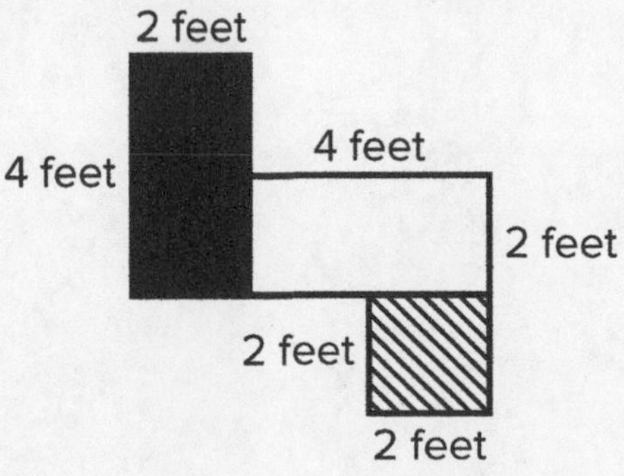

area = _____ square feet

2. Tristan is building a pen with an area of 10 square feet. His dad wants him to increase its length to 8 feet. How will its area change?

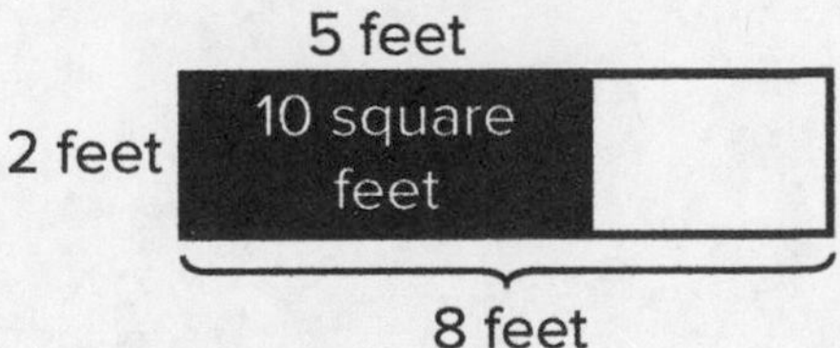

It will increase by _____ square feet. The new area will be _____ square feet.

Solve Area Problems

Name ______________________________

Solve. Show your work.

1. Each roll of wallpaper has an area of 50 square feet. How many rolls of wallpaper would you need to cover four walls if their dimensions are 5 feet by 5 feet?

2. How many rolls of wallpaper with an area of 36 square feet would you need to cover four walls if their dimensions are 6 feet by 6 feet?

Partition Shapes into Equal Parts

Name ______________________________

Review

You can draw lines to partition a shape into **equal parts.**
You can partition the shape in different ways.

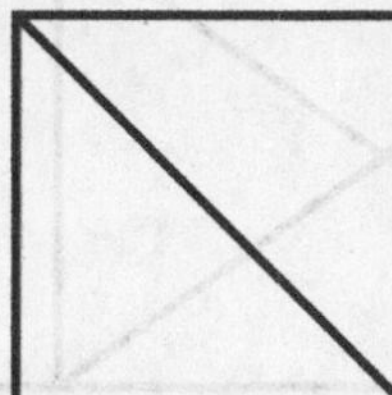

two halves
two equal parts

Draw lines to partition each shape into equal parts three different ways.

1. fourths

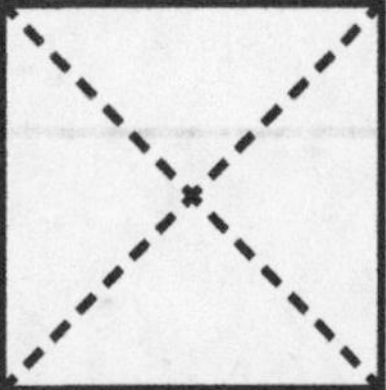
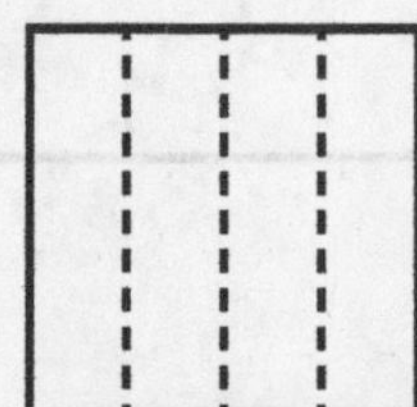

2. sixths

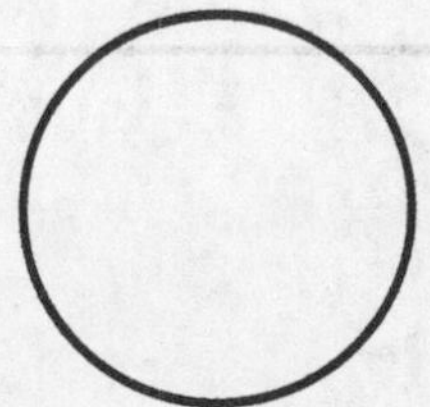

3. eighths

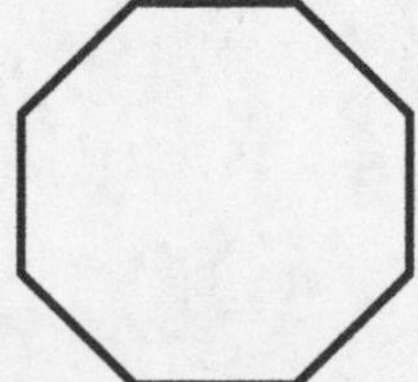

Partition Shapes into Equal Parts

Name ____

Rena partitioned the rectangle into 8 equal parts. The parts are not all the same shape but each part is one-eighth of the whole.

Draw lines to partition a rectangle into 8 equal parts are not all the same shape in a different way than Rena.

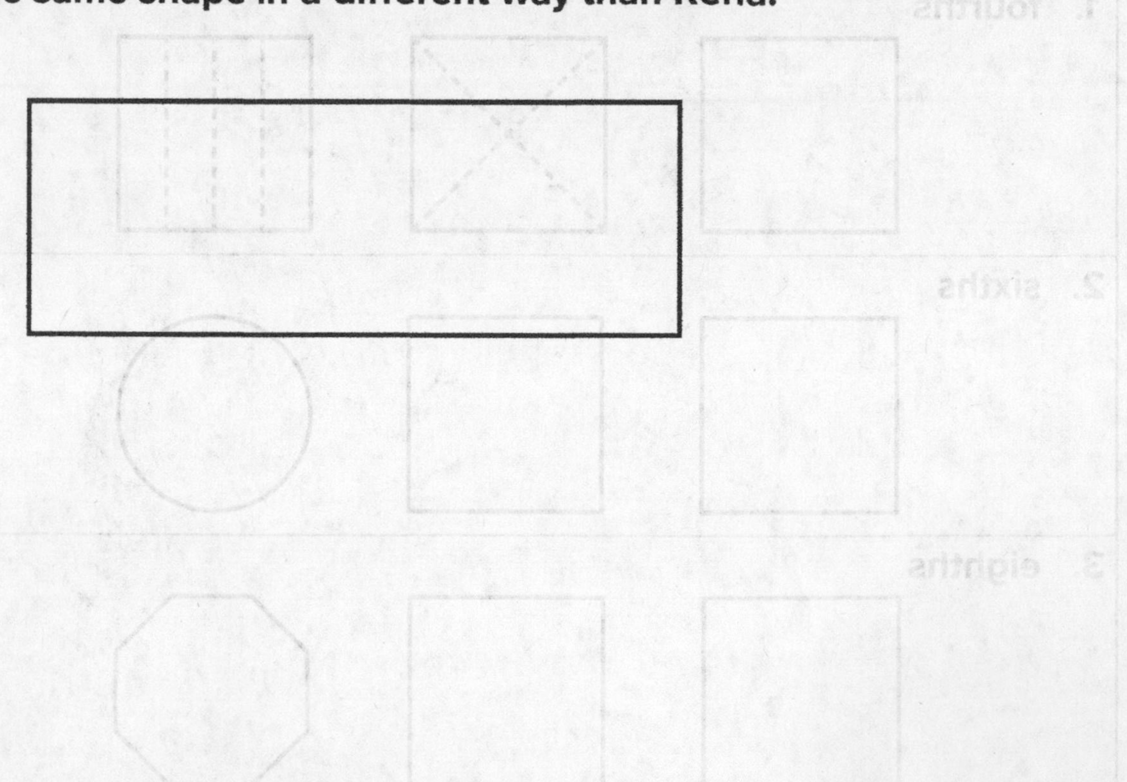

Understand Fractions

Name ______________________________

Review

You can partition a whole shape into **equal parts.**

This whole shape has four equal parts.

Each part represents **one-fourth.**

You can write **one-fourth** as a **fraction.**

Write $\frac{1}{4}$ ← numerator ← denominator

Write the fraction represented by each equal part.

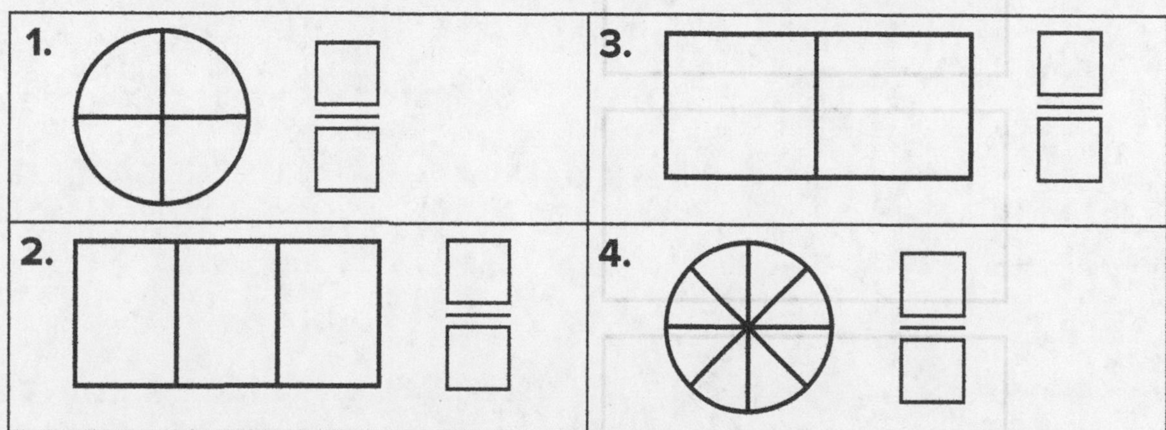

Write the fraction represented by the shaded part.

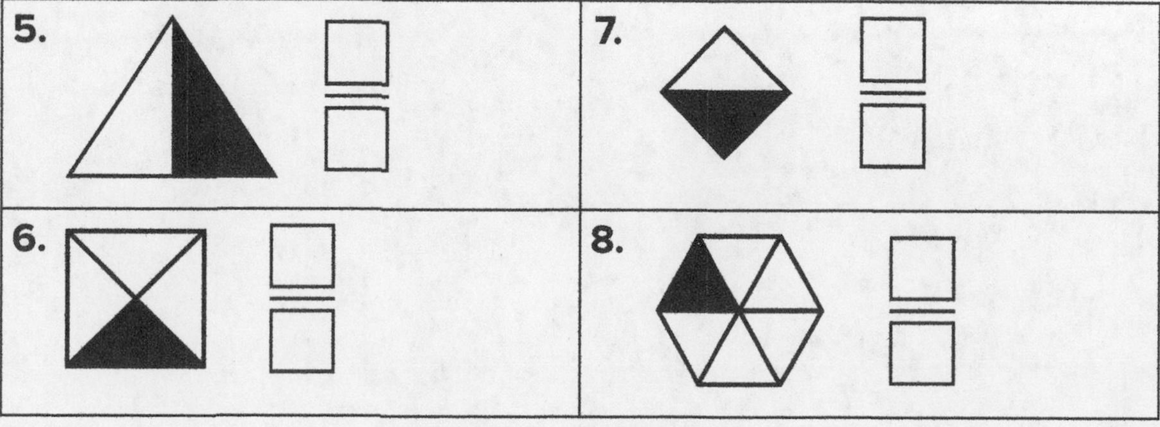

Understand Fractions

Name ________________________________

Ravi says $\frac{1}{2}$ of this figure is shaded? What would you say to Ravi? How else can the rectangle be divided evenly with the same amount shaded? What fraction represents that amount? Show your work.

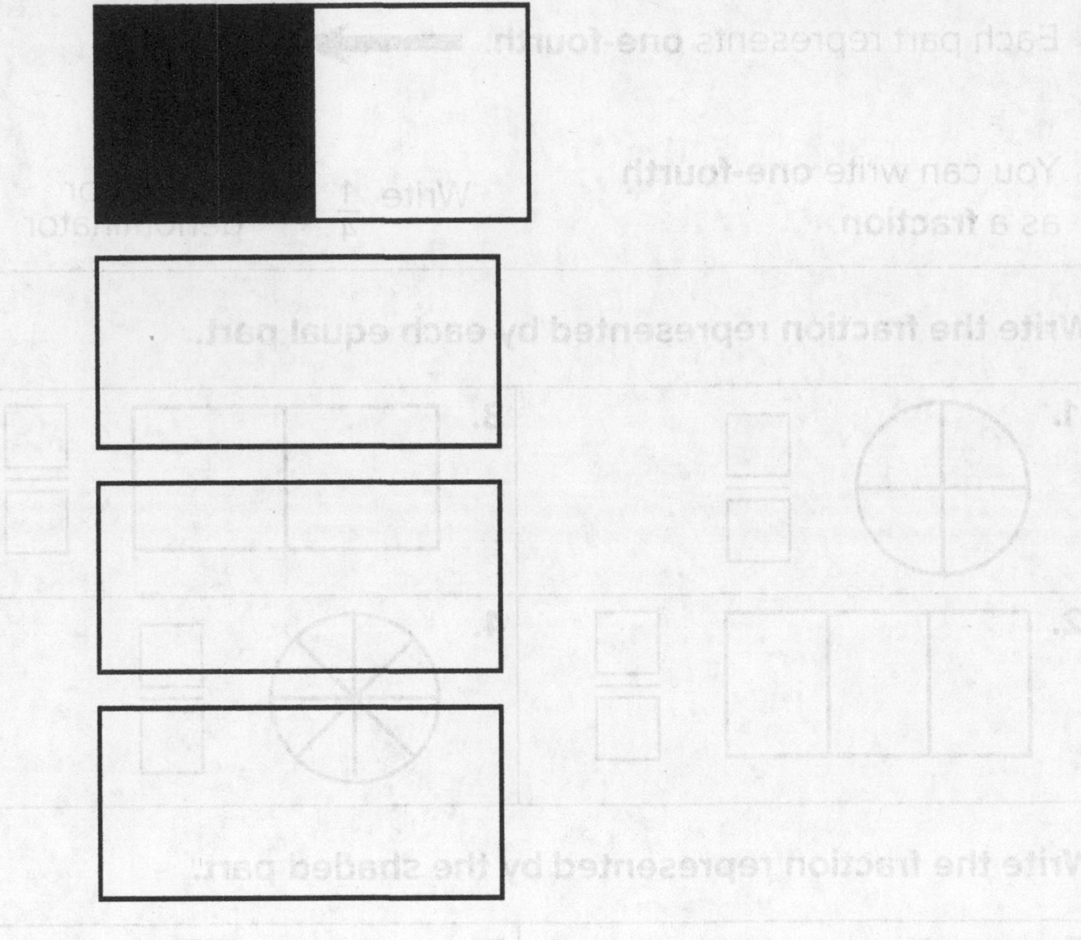

Represent Fractions on a Number Line

Name ______________________________

Review

You can partition a number line into equal parts like you partition a shape.

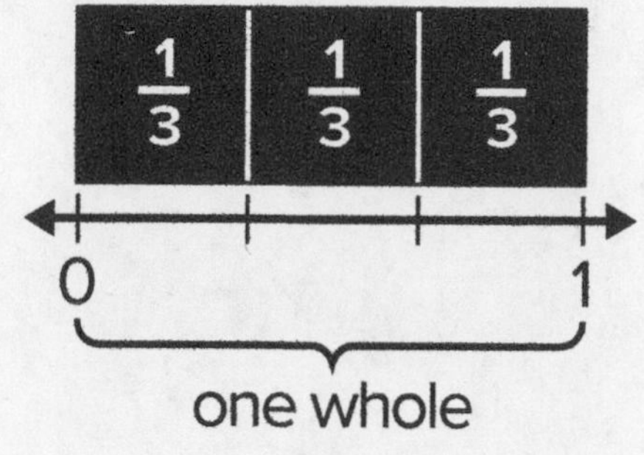

Each interval from 0 to 1 represents a fraction.

Partition the number line into 3 equal parts.

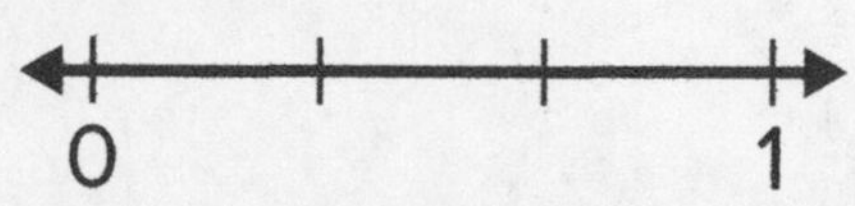

Count unit fraction intervals to name the other fractions

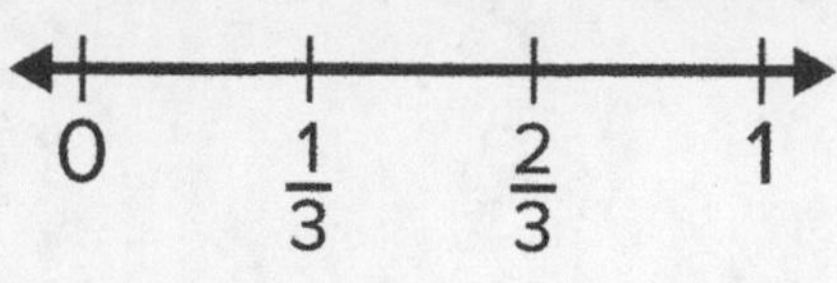

Write the fractions to complete the number lines.

1. 0 $\quad \frac{1}{4} \quad \frac{\square}{\square} \quad \frac{\square}{\square}$ $\quad$ 1

2.

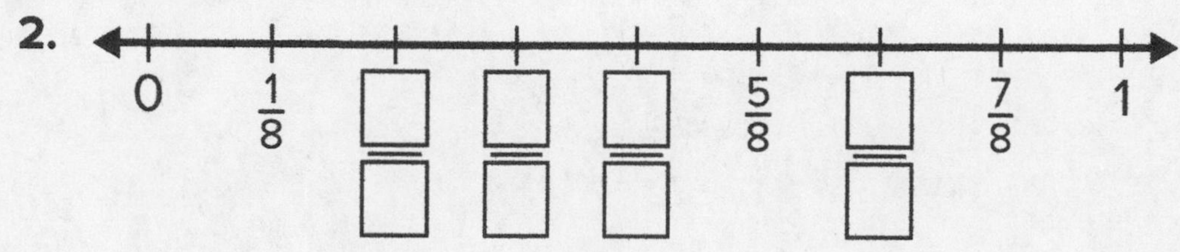

Represent Fractions on a Number Line

Name ______________________________

1. Marcy and Sam are swimming in a one-mile race. Marcy has completed $\frac{1}{2}$ of the race. Sam has completed $\frac{3}{8}$ of the race. Who is winning? Draw number lines to explain.

2. Who would be winning if Sam had completed $\frac{4}{6}$ of the mile? Draw a number line to explain.

Represent One Whole as a Fraction

Name ______________________________

Review

You can partition and shade a shape or use a number line to see how many parts equals one.

four-fourths = 1 whole You can write four-fourths as a fraction that equals 1.

$\frac{4}{4}$

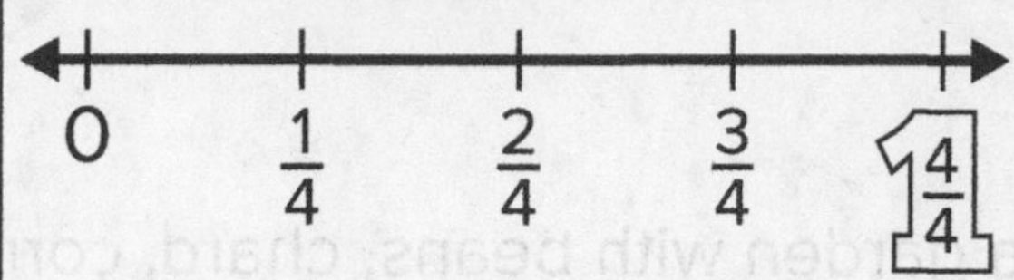

$\frac{4}{4}$ and 1 are the same point on the number line.

A fraction is equal to 1 when the numerator and the denominator represent the same number of parts.

Write the fraction that represents the shaded part.

1.

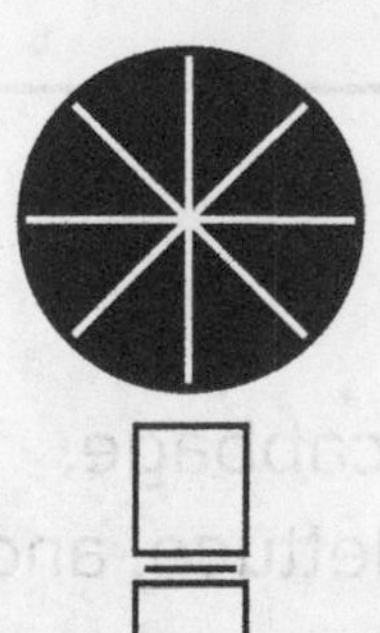

2.

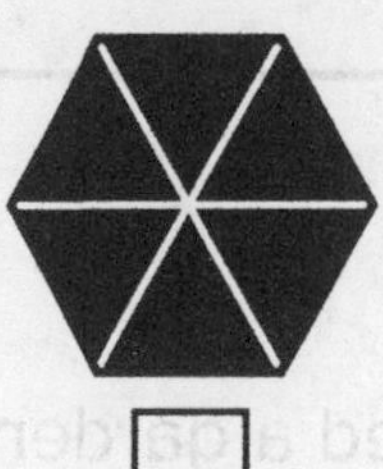

Label the number line using fractions. Write the fraction that represents 1.

3.

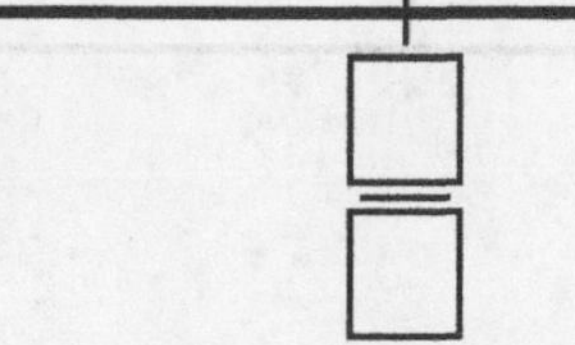

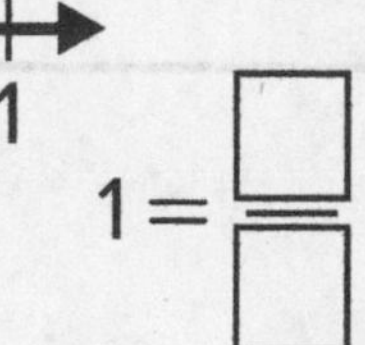

Represent One Whole as a Fraction

Name ______________________________

Label the garden with each vegetable or fruit and write a fraction to represent the number of parts with planted vegetables or fruits. Explain your answer.

1. Carina planted a garden with carrots, cucumbers, peas, kale, radishes, and tomatoes. Each part has one vegetable in it.

2. David's grandfather planted a garden with beans, chard, corn, watermelon, cantaloupe, summer squash, corn, and tomatoes. Each part has one vegetable or fruit in it.

3. Louella planted a garden with peppers, beans, cabbage, cucumbers, onions, zucchini, carrots, radishes, lettuce, and beets. Each part has one vegetable in it.

Represent Whole Numbers as Fractions

Name ______________________________

Review

You can write whole numbers as fractions

Example 1

each shape = 1 whole

4 wholes = $\frac{4}{1}$

Example 2

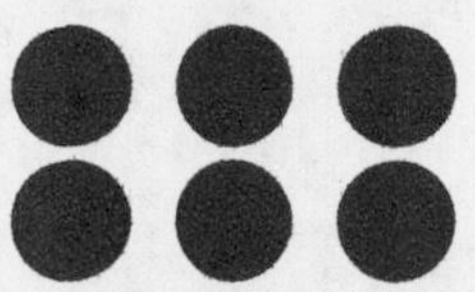

6 wholes = $\frac{6}{1}$

Write the fraction that represents the whole number.

1. 2 = $\frac{\square}{\square}$

2. 

7 = $\frac{\square}{\square}$

3. 

8 = $\frac{\square}{\square}$

4. 

5 = $\frac{\square}{\square}$

5. 

9 = $\frac{\square}{\square}$

6. 

10 = $\frac{\square}{\square}$

7. Circle the fractions that are equal to a whole number.

$\frac{9}{1}$ $\frac{2}{4}$ $\frac{3}{6}$

$\frac{6}{8}$ $\frac{5}{1}$ $\frac{8}{1}$

Represent Whole Numbers as Fractions

Name ______________________________

1. Four friends had 12 small oranges to take on a long hike. Amy and Brooke each ate 2 oranges. Together Mark and David ate 4 times as many oranges as Amy and Brooke.

How can you express the number of oranges Mark and David ate as a fraction? Explain your answer.

2. Write a fraction to represent the number of oranges Mark and David would have eaten if the children had 18 small oranges, if Brooke and Amy each ate 3 oranges instead of 2. Explain your answer.

Represent a Fraction Greater Than One on a Number Line

Name ______________________________

Review

You can label a number line with **fractions greater than 1.**

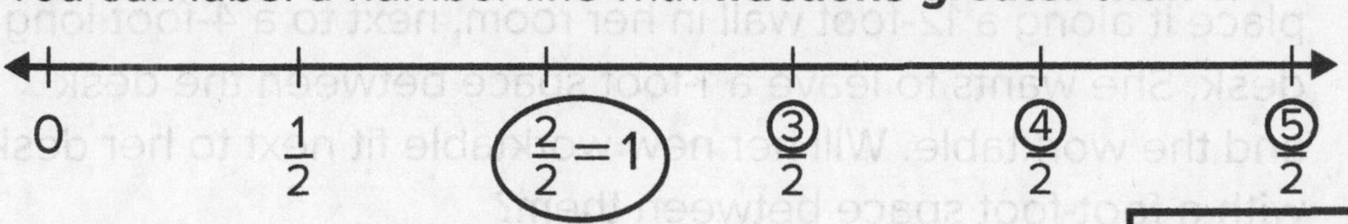

The fractions greater than 1 are $\frac{3}{2}$, $\frac{4}{2}$, and $\frac{5}{2}$.

HINT:
Fractions greater than 1 have a numerator that is greater than the denominator.

Write the fractions to complete the number lines. Circle the fractions great than 1.

1.

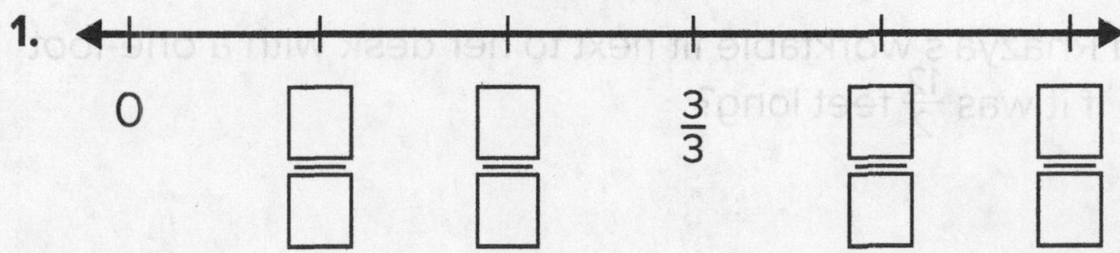

2.

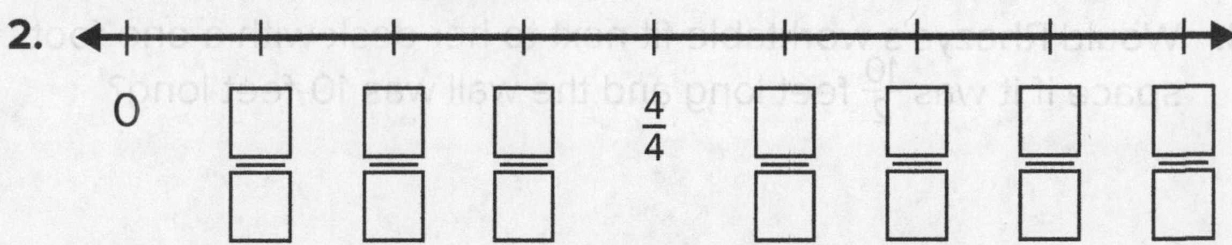

Represent a Fraction Greater Than One on a Number Line

Name ______________________________

Solve. Draw a number line to justify your answer.

1. Rhayza bought a worktable that is $\frac{16}{2}$ feet long. She wants to place it along a 12-foot wall in her room, next to a 4-foot long desk. She wants to leave a 1-foot space between the desk and the worktable. Will her new worktable fit next to her desk with a foot-foot space between them?

2. Would Rhazya's worktable fit next to her desk with a one-foot space if it was $\frac{12}{2}$ feet long?

3. Would Rhazya's worktable fit next to her desk with a one-foot space if it was $\frac{10}{2}$ feet long and the wall was 10-feet long?

Understand Equivalent Fractions

Name ______________________________

Review

Fractions that represent the same part of a whole are **equivalent**.

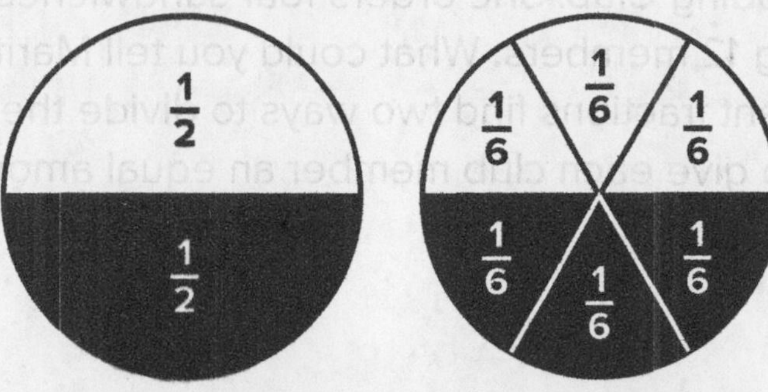

$\frac{1}{2}$ and $\frac{3}{6}$ are **equivalent** fractions
They shade the same amount of the whole.

Shade the model to show the equivalent fraction.

1. 

2. 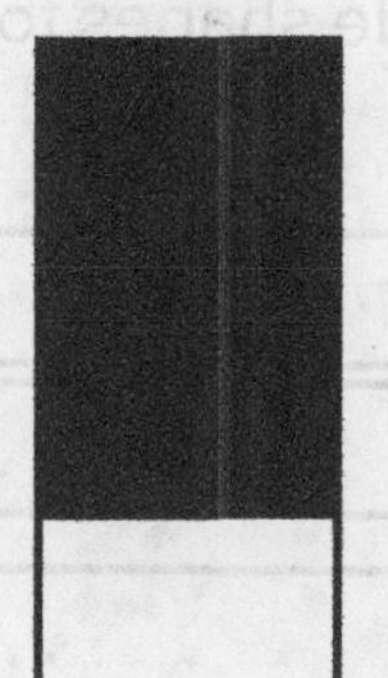 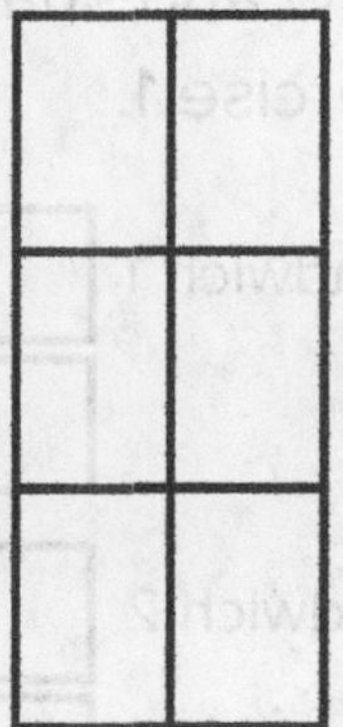

Use the fraction models to determine if the fractions are equivalent. Write *equivalent* or *not equivalent*.

3. $\frac{1}{2}$ and $\frac{2}{4}$

4. $\frac{1}{4}$ and $\frac{2}{6}$

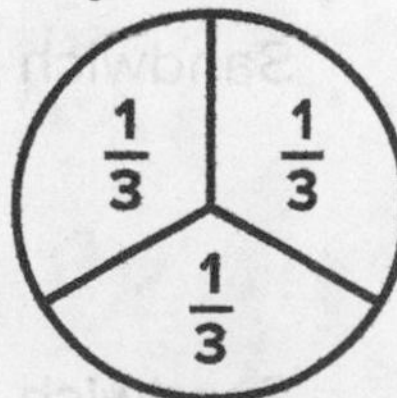

5. $\frac{1}{3}$ and $\frac{3}{4}$

6. $\frac{3}{3}$ and $\frac{6}{6}$

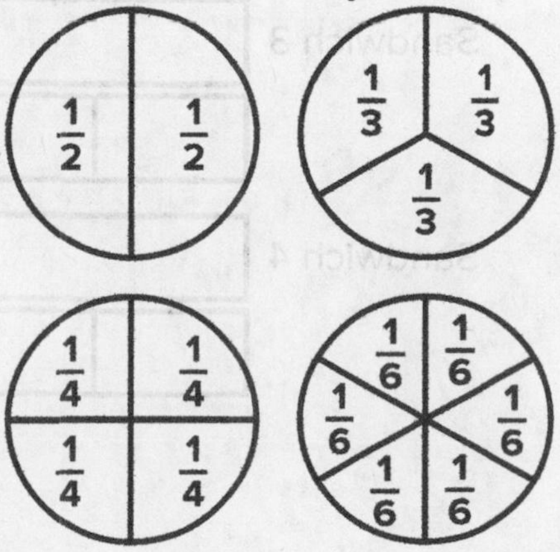

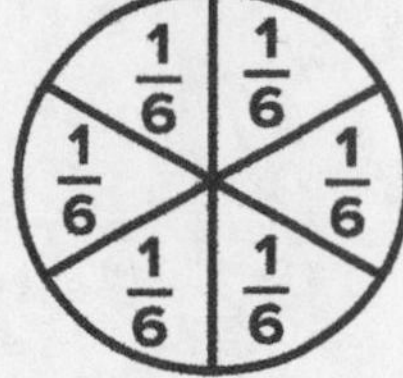

Lesson **8-1 • Extend Thinking**

Understand Equivalent Fractions

Name ______________________________

Solve. Explain your work.

1. Marita is in charge of ordering submarine sandwiches for her afterschool Coding Club. She orders four sandwiches to share equally among 12 members. What could you tell Marita about using equivalent fractions find two ways to divide the four sandwiches to give each club member an equal amount?

2. Draw and shade shapes to justify your answer to Exercise 1.

Sandwich 1

Sandwich 2

Sandwich 3

Sandwich 4

Represent Equivalent Fractions

Name ______________________

Review

You can partition a whole into equal parts in different ways to represent equivalent fractions.

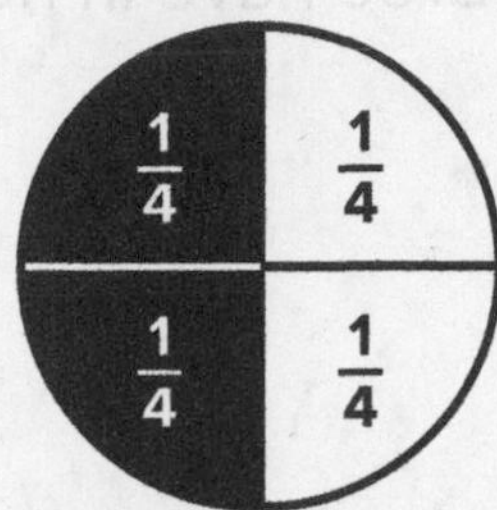

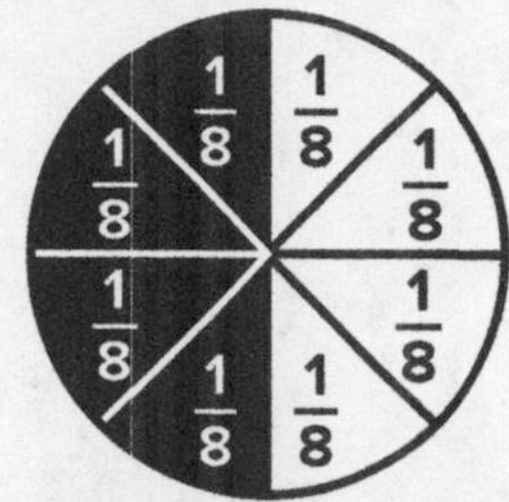

The fractions $\frac{2}{4}$ and $\frac{4}{8}$ shade the same part of the same-size whole. They are equivalent fractions. $\frac{2}{4} = \frac{4}{8}$.

Shade the model and complete the fraction to show the fractions are equivalent

1.

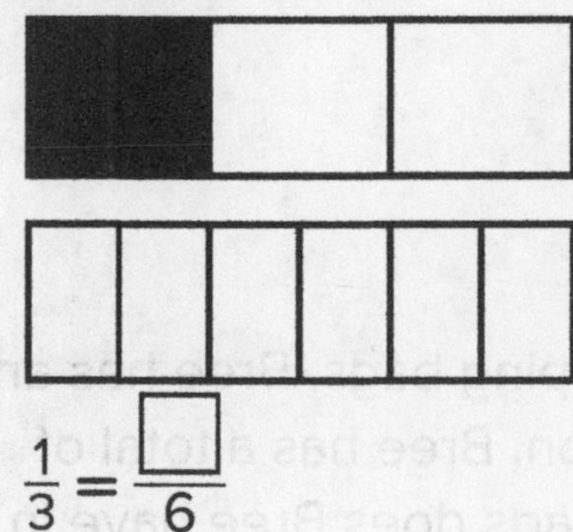

$\frac{1}{3} = \frac{\square}{6}$

2.

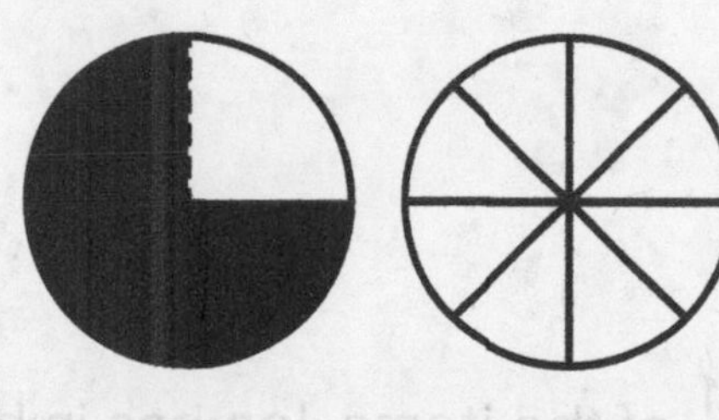

$\frac{3}{4} = \frac{\square}{8}$

Draw fraction models to complete the equation.

3. $\frac{1}{\square} = \frac{3}{6}$

4. $\frac{\square}{3} = \frac{4}{6}$

5. $\frac{6}{6} = \frac{\square}{8}$

6. $\frac{1}{\square} = \frac{4}{8}$

Represent Equivalent Fractions

Name ____________________

Solve. Draw and shade fraction bars to explain your work.

1. $\frac{1}{4}$ of the items Jon has in his tent are lanterns. Bree has an equivalent fraction of lanterns as Jon. Bree has a total of 8 items in her tent. How many lanterns does Bree have in her tent?

2. $\frac{1}{3}$ of the items Jon has outside of his tent are camp chairs. Bree has an equivalent fraction of camp chairs as Jon. Bree has a total of 6 items outside of her tent. How many camp chairs does Bree have outside of her tent?

3. $\frac{1}{2}$ of the items Jon has in his tent are sleeping bags. Bree has an equivalent fraction of sleeping bags as Jon. Bree has a total of 8 items in her tent. How many sleeping bags does Bree have in her tent?

Represent Equivalent Fractions on a Number Line

Name ____

Review

Fractions at the same point on a number line represent equivalent fractions.

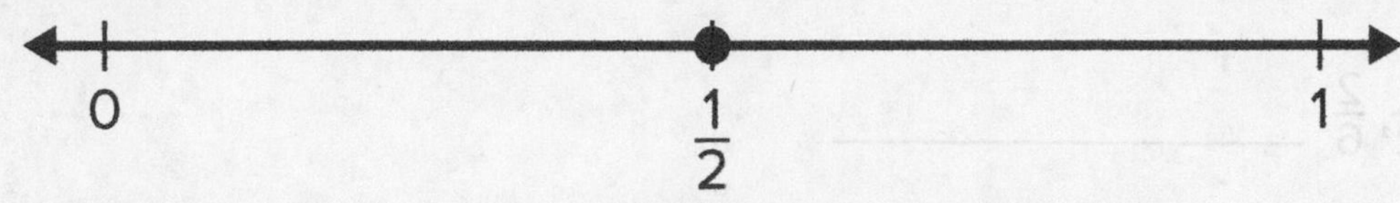

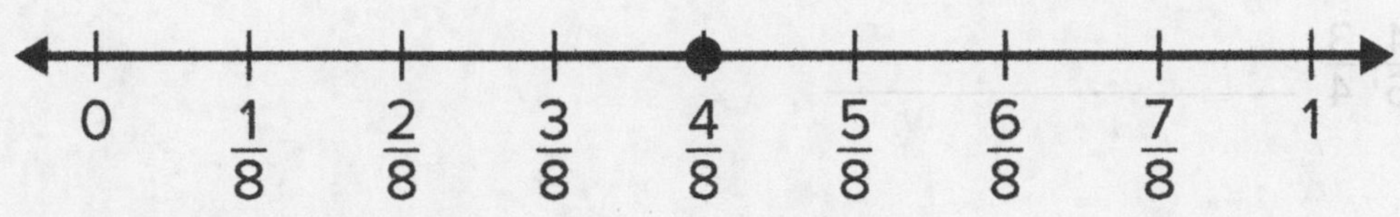

The points at $\frac{1}{2}$ and $\frac{4}{8}$ are equivalent because they are the same distance from 0.

Use the points on the number lines to name the equivalent fractions.

1.

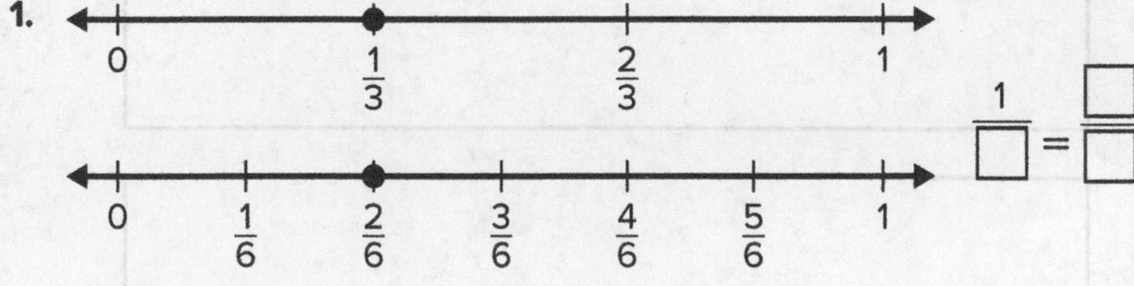

$\frac{1}{\square} = \frac{\square}{\square}$

2.

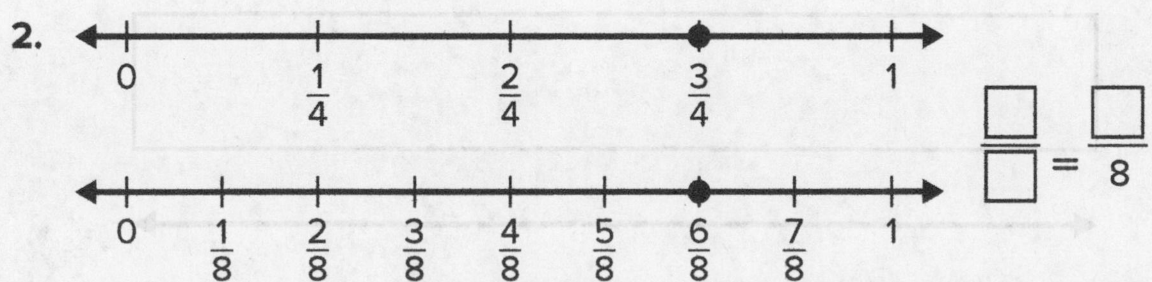

$\frac{\square}{\square} = \frac{\square}{8}$

Represent Equivalent Fractions on a Number Line

Name ______________________________

Write equivalent or not equivalent. Use different representations to justify the answers.

1. $\frac{4}{8}, \frac{1}{2}, \frac{3}{6}$ ____________________

2. $\frac{1}{3}, \frac{1}{2}, \frac{2}{6}$ ____________________

3. $\frac{2}{3}, \frac{4}{6}, \frac{3}{4}$ ____________________

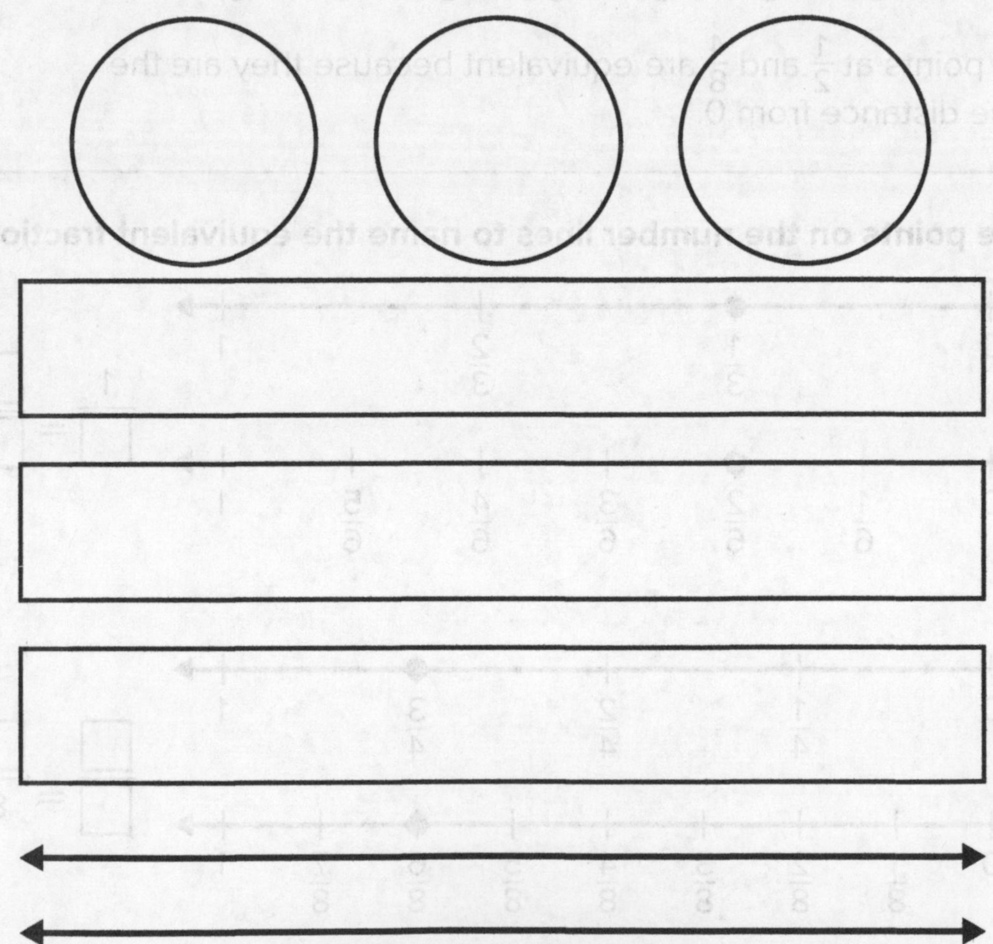

Lesson 8-4 • Reinforce Understanding

Understand Fractions of Different Wholes

Name ______________________________

Review

You can tell if fraction models represent the same amount only if the wholes are the same shape and size.

same amount

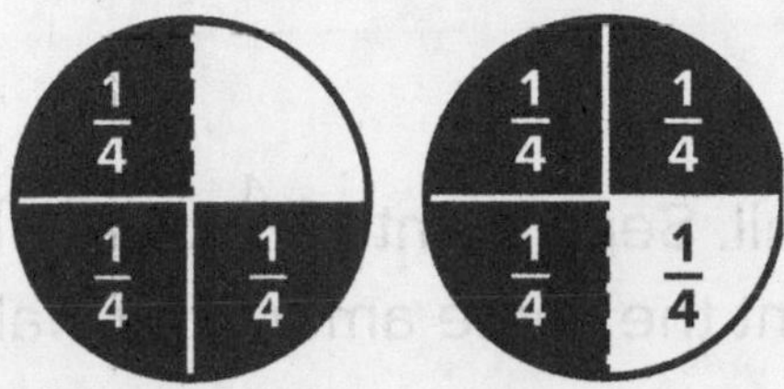

same shape and same size

not the same amount

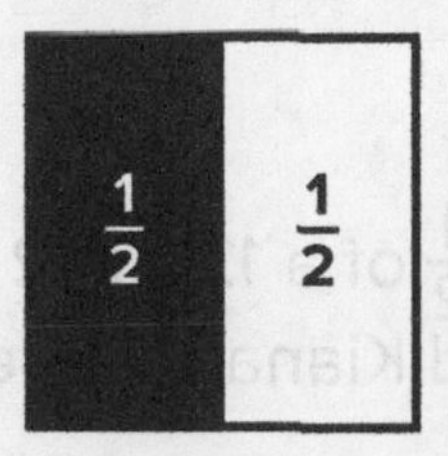

same shape but different size

Write *yes* or *no* to tell if the parts are equivalent.

1. ____________

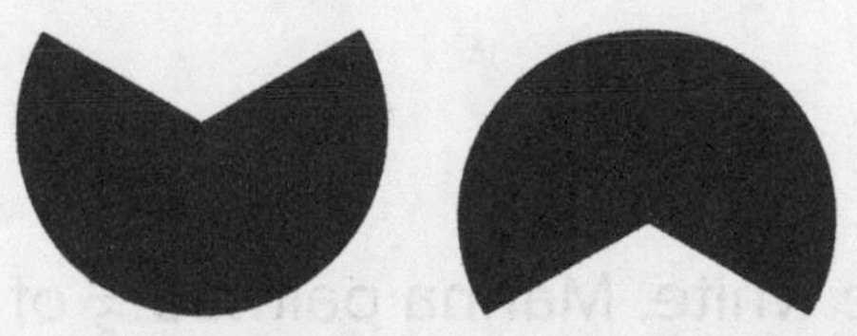

2. ____________

Draw a picture to match the statement.

3. Two models of $\frac{1}{6}$ that represent the same amount.

4. Two models of $\frac{1}{8}$ that do not represent the same amount.

Understand Fractions of Different Wholes

Name ______________________________

Solve. Explain your answer.

1. Kai painted $\frac{1}{3}$ of a 15 foot piece of wood. Marita painted $\frac{1}{3}$ of a 16 inch piece of wood. Did Kai and Marita paint the same amount of wood?

2. Kiana painted $\frac{1}{2}$ of a 12 by 12 foot wall. Sean painted $\frac{4}{8}$ of a 12 by 12 foot wall. Did Kiana and Sean paint the same amount of wall?

3. Dawna decorated $\frac{1}{4}$ of a 10 by 12 foot piece of plywood. Wally decorated $\frac{2}{8}$ of a 10 by 10 inch piece of plywood. Did Dawna and Wally decorate the same amount of plywood?

4. Deanna painted $\frac{2}{6}$ of an 8 foot fence white. Marina painted $\frac{2}{3}$ of the fence white. Did Deanna and Marina paint the same amount of the fence?

5. Ravi decorated $\frac{3}{6}$ of a 5 by 3 foot canvas. Sheena decorated $\frac{1}{2}$ of a 5 by 5 inch canvas. Did Ravi and Sheena decorate the same amount of canvas?

Lesson **8-5 • Reinforce Understanding**

Compare Fractions with the Same Denominator

Name ______________________________

Review

You can use the size of the numerator to compare fractions with the same denominator.

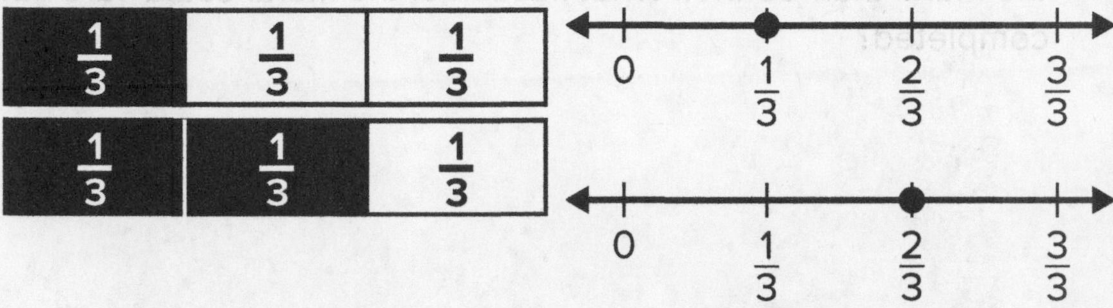

$\frac{2}{3}$ is **greater than** $\frac{1}{3}$

$\frac{2}{3} \boxed{>} \frac{1}{3}$

$\frac{1}{3}$ is **less than** $\frac{2}{3}$

$\frac{1}{3} \boxed{<} \frac{2}{3}$

When comparing fractions with the same denominator, the fraction with the greater numerator is greater.

Write >, <, or = to make the comparison true. Shade the fraction tiles to justify your reasoning

1. $\frac{3}{6} \square \frac{5}{6}$

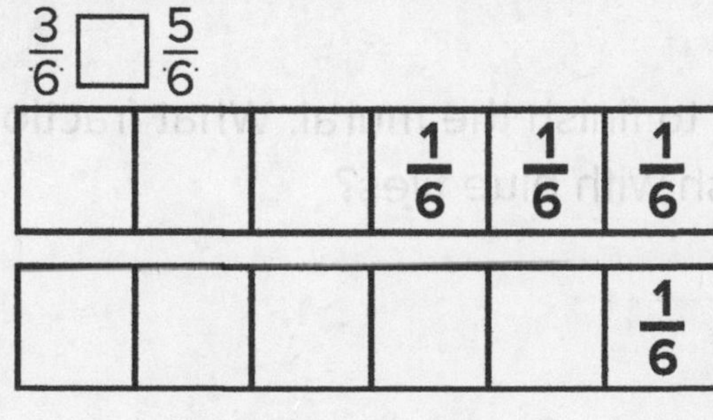

2. $\frac{2}{4} \square \frac{3}{4}$

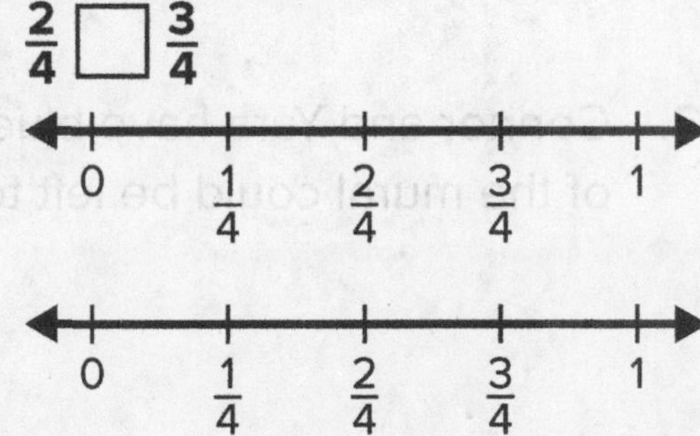

Compare Fractions with the Same Denominator

Name ______________________________

Solve. Explain your reasoning.

1. Connor and Yara are making a mosaic mural in their classroom. Connor has completed $\frac{1}{4}$ of the mural. Yara has completed more of the mural than Connor. What fraction of the mural could Yara have completed?

2. Connor used red and orange tiles to make $\frac{3}{6}$ of the mural. Yara used yellow and green tiles to make more of the mural than Connor. What fraction of the mural could Yara have made with the yellow and green tiles?

3. Connor and Yara have blue tiles to finish the mural. What fraction of the mural could be left to finish with blue tiles?

Compare Fractions with the Same Numerator

Name ____________________

Review

You can use the size of the denominator to compare fractions with the same numerator.

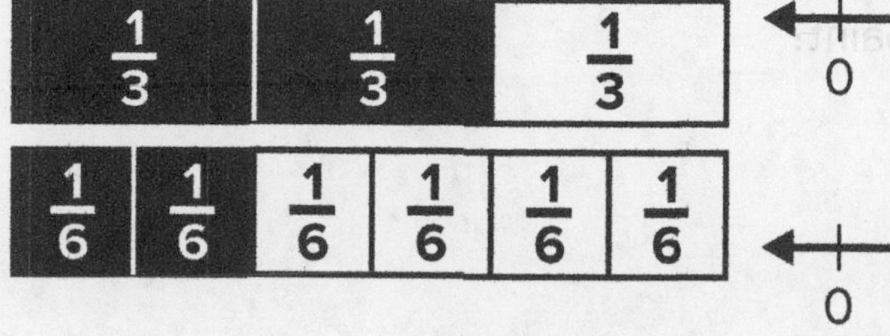

$\frac{2}{3}$ is **greater than** $\frac{2}{6}$

$\frac{2}{3} > \frac{2}{6}$

$\frac{2}{6}$ is **less than** $\frac{2}{3}$

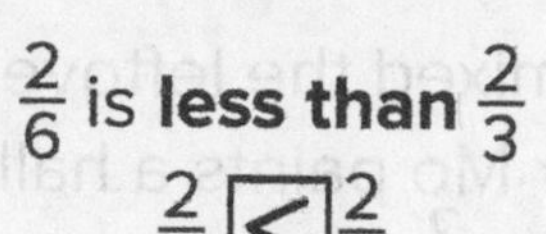

$\frac{2}{6} < \frac{2}{3}$

When comparing fractions with the same numerator, the fraction with the lesser denominator is greater.

Write >, <, or = to make the comparison true. Shade the fraction tiles to justify your reasoning

1. $\frac{3}{6}$ ☐ $\frac{3}{4}$

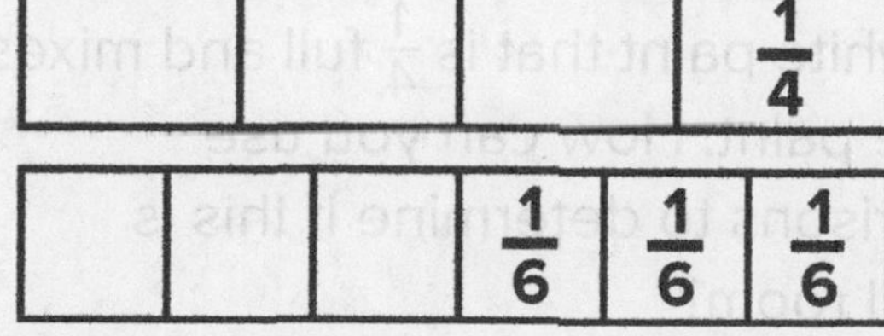

2. $\frac{2}{6}$ ☐ $\frac{2}{4}$

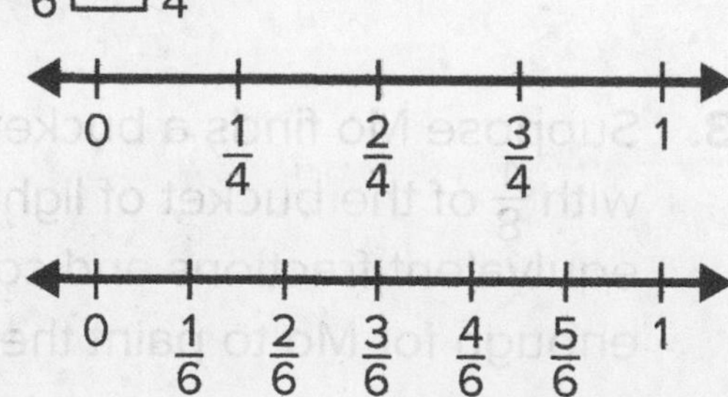

Compare Fractions with the Same Numerator

Name ______________________________

Solve. Explain your reasoning.

1. After a painting a room, a $\frac{1}{2}$ bucket of white paint and a $\frac{1}{8}$ bucket of blue paint are left. The buckets are the same size. How can you use equivalent fractions and comparisons to determine which bucket has less paint?

2. Mo mixed the leftover paint to make $\frac{5}{8}$ bucket of light blue paint. After Mo paints a hallway, there is $\frac{2}{4}$ of the bucket of paint left. He needs $\frac{2}{8}$ of a bucket to paint a small room. How can you use equivalent fractions and comparisons to determine if Mo has enough paint left to paint the small room?

3. Suppose Mo finds a bucket of white paint that is $\frac{1}{4}$ full and mixes it with $\frac{1}{8}$ of the bucket of light blue paint. How can you use equivalent fractions and comparisons to determine if this is enough for Mo to paint the small room?

Compare Fractions

Name ______________________________

Review

You can use fraction models and number lines to justify comparisons between two fractions. Compare $\frac{1}{3}$, $\frac{2}{3}$, and $\frac{2}{6}$.

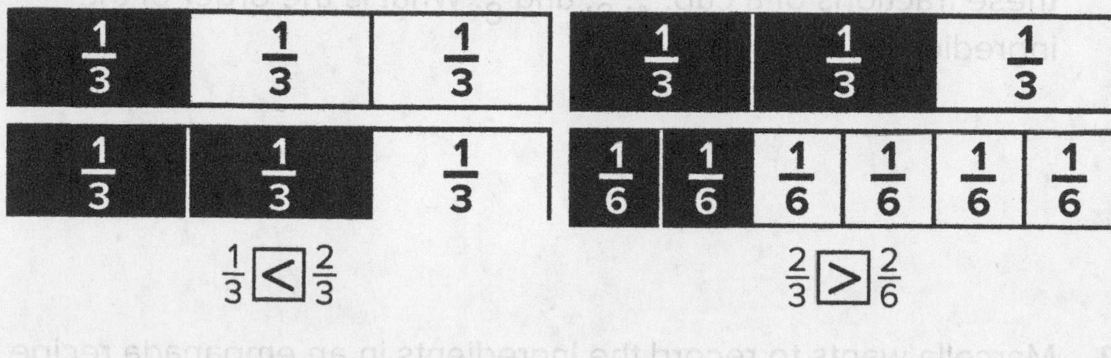

Write >, <, or = to make the comparison true. Draw a fraction model or two number lines to justify the answer.

1. $\frac{1}{6}$ ☐ $\frac{3}{6}$

2. $\frac{3}{4}$ ☐ $\frac{3}{6}$

3. $\frac{3}{8}$ ☐ $\frac{3}{6}$

Circle the comparisons that are true.

4. $\frac{2}{4} = \frac{4}{8}$ $\qquad \frac{4}{6} > \frac{6}{4}$ $\qquad \frac{2}{6} < \frac{4}{6}$ $\qquad \frac{4}{7} > \frac{4}{8}$

Compare Fractions

Name ______________________________

Solve. Draw a representation to justify your answer.

1. Janine wants to organize the ingredients in a pretzel recipe from the least to greatest amount. The ingredients are measured in these fractions of a cup: $\frac{3}{4}$, $\frac{3}{8}$, and $\frac{5}{8}$. What is the order of the ingredients?

2. Marcella wants to record the ingredients in an empanada recipe in order from the greatest to least amount. The ingredients are measured in these fractions of a cup: $\frac{2}{4}$, $\frac{5}{6}$, and $\frac{2}{3}$. What is the order of the ingredients?

3. Carol wants to write the ingredients in a roti recipe in order from the least to greatest amount. The ingredients are measured in these fractions of a cup: $\frac{2}{8}$, $\frac{3}{4}$, and $\frac{2}{3}$. What is the order of the ingredients?

Use Multiplication to Solve Division Equations

Name ________________________________

HINT:
The quotient is the unknown factor.

Review

You can use the relationship between multiplication and division to represent a division equation as an unknown-factor problem.

Step 1 Rewrite **12 ÷ 4 = ?** as an unknown factor problem.

? × 4 = 12

Step 2 Use an array to help you find the unknown factor.

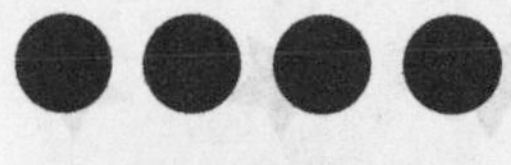

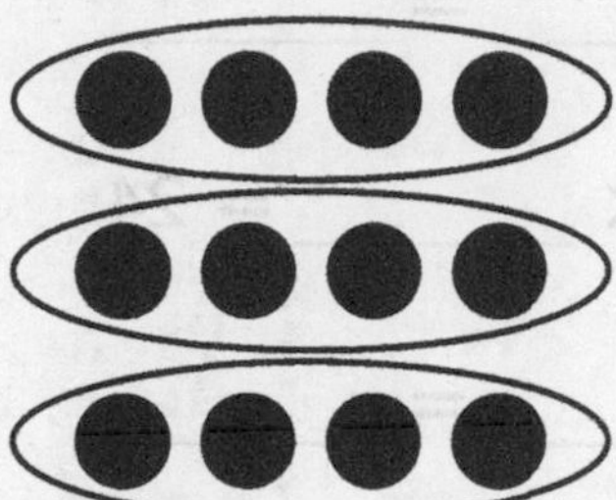

3 × 4 = 12

12 ÷ 4 = **3**

Complete the equations and draw an array.

1. 24 ÷ 6 = ________

________ × 6 = 24

2. 27 ÷ 9 = ________

________ × 9 = 27

Write an unknown-factor problem for the division equation.

3. 42 ÷ 6 = ?

________ × ? = ________

4. 25 ÷ ? = 5

? × ________ = ________

5. 28 ÷ 7 = ?

____

6. 16 ÷ ? = 8

____

Use Multiplication to Solve Division Equations

Name ______________________________

Solve.

1. Mario drew an array and wrote parts of the 4 equations it represents. Complete the equations for his array and explain how it represents 2 multiplication and 2 division equations.

______ × ______ = 24

24 ÷ ______ = ______

______ × ______ = 24

24 ÷ ______ = ______

2. Draw your own array and write the 4 equations it represents.

______ × ______ = ______

______ ÷ ______ = ______

______ × ______ = ______

______ ÷ ______ = ______

Divide by 2

Name ____

Review

You can use the relationship between multiplication and division to divide by 2.

You can draw a diagram to help you find the unknown.	You can draw equal groups to help you find the unknown.
5 | 5 ; ? $10 \div 2 = \mathbf{5}$ $2 \times \mathbf{5} = 10$	 $12 \div 2 = \mathbf{6}$ $2 \times \mathbf{6} = 12$

Circle the multiplication equations that can help you find the missing quotient for $18 \div 2 = ?$

1. $18 \times 2 = ?$ $\quad ? \times 2 = 18$ $\quad 18 = 2 \times ?$ $\quad ? \times 2 = 9$

Complete the equation to make it true.

2. $6 \div 2 =$ ________

3. ________ $= 8 \div 2$

4. ________ $= 20 \div 2$

5. $16 \div 2 =$ ________

Draw a model and write a related multiplication equation to complete the division equation.

6. $14 \div 2 = ?$

Divide by 2

Name ______________________________

Spin the pointer of an 8-part spinner. Use the number the pointer lands on as the unknown factor. Use it to complete the multiplication and division equations.

1. ______ × 2 = ______

______ ÷ 2 = ______

2 × ______ = ______

______ ÷ 2 = ______

2. ______ × 2 = ______

______ ÷ 2 = ______

2 × ______ = ______

______ ÷ 2 = ______

3. ______ × 2 = ______

______ ÷ 2 = ______

2 × ______ = ______

______ ÷ 2 = ______

4. ______ × 2 = ______

______ ÷ 2 = ______

2 × ______ = ______

______ ÷ 2 = ______

5. ______ × 2 = ______

______ ÷ 2 = ______

2 × ______ = ______

______ ÷ 2 = ______

6. ______ × 2 = ______

______ ÷ 2 = ______

2 × ______ = ______

______ ÷ 2 = ______

7. ______ × 2 = ______

______ ÷ 2 = ______

2 × ______ = ______

______ ÷ 2 = ______

8. ______ × 2 = ______

______ ÷ 2 = ______

2 × ______ = ______

______ ÷ 2 = ______

Divide by 5 and 10

Name ________________________________

Review

You can complete division facts with 5 and 10 by recalling related multiplication facts.

Think: what number multiplied by 5 equals 15?	**Think:** what number multiplied by 10 equals 20?
$15 \div 5 = \mathbf{?}$ $5 \times \mathbf{?} = 15$	$20 \div 10 = \mathbf{?}$ $10 \times \mathbf{?} = 20$
$5 \times \mathbf{3} = 15$	$10 \times \mathbf{2} = 20$
$15 \div 5 = \mathbf{3}$	$20 \div 10 = \mathbf{2}$

Circle the multiplication equations that can help you find the missing quotients for $20 \div 5 = ?$

1. $20 \times 5 = ?$ $? \times 5 = 20$ $20 = 5 \times ?$ $? \times 5 = 4$

Complete the equation to make it true.

2. $30 \div 5 =$ ______   **3.** ______ $= 70 \div 10$

4. ______ $= 45 \div 5$ **5.** $50 \div 10 =$ ______

Draw a line to the related expression.

6. $50 \div 10$ 4×5

$20 \div 5$ 10×5

Divide by 5 and 10

Name ______________________________

Write a related expression.

1. 50 ÷ 5 ________ ________ ________

2. 5 × 4 ________ ________ ________

3. 40 ÷ 10 ________ ________ ________

4. 35 ÷ 5 ________ ________ ________

5. 5 × 8 ________ ________ ________

6. 10 × 10 ________ ________ ________

7. 70 ÷ 10 ________ ________ ________

8. 10 × 6 ________ ________ ________

9. 90 ÷ 10 ________ ________ ________

10. 5 × 6 ________ ________ ________

Understand Division with 1 and 0

Name ____

Review

You can use patterns and rules to recall division facts with 1 and 0.

Divide and Multiply by 1

If you divide any number by 1 the quotient is that number.	If you multiply any number by 1 the product is that number.
$\mathbf{1 \div 1 = 1}$ $\mathbf{2 \div 1 = 2}$ $\mathbf{3 \div 1 = 3}$	$\mathbf{1 \times 1 = 1}$ $\mathbf{2 \times 1 = 2}$ $\mathbf{3 \times 1 = 3}$

Divide and Multiply by 0

If you divide 0 by a number the quotient is always 0.	If you multiply any number by 0 the product is always 0.
$\mathbf{0 \div 1 = 0}$	$\mathbf{1 \times 0 = 0}$ $\mathbf{2 \times 0 = 0}$

You cannot divide by 0 since multiplying by 0 = 0.

$3 \div 0 = \mathbf{X}$

Complete the equation to make it true. Cross out equations that cannot be solved.

1. $3 \div$ ________ $= 1$

2. $0 =$ ________ $\div 0$

3. $4 = 4 \div$ ________

4. $6 \div 1 =$ ________

5. ________ $= 5 \div 0$

6. $2 \div$ ________ $= 1$

7. ________ $= 8 \div 8$

8. ________ $= 0 \div 9$

9. ________ $= 6 \div 6$

10. ________ $= 5 \div 5$

Understand Division with 1 and 0

Name ______________________________

Solve. Explain your work.

1. Mark checks out board games from the Library of Things. He plays 1 game per week. How many weeks will it take Mark to play all of the games?

2. Mark's sister checks knitting needles out from the Library of Things. She knits 1 headscarf per month. How many headscarves can she knit in 3 months?

3. Mark's aunt checks out a 3D printing pen from the Library of Things. She makes 1 toy dinosaur per day. How many days will it take her to make toy dinosaurs for her 8 nephews?

4. Mark's dad checks out a hand lettering kit from the Library of Things. It takes 1 hour to hand letter 1 invitation. He wants to hand letter 0 invitations. How many hours will it take him?

Divide by 3 and 6

Name ____

Review

You can use related multiplication facts to divide by 3 and 6.

Divide by 3

21 ÷ 3 = **?**

Think: 3 × **?** = 27

3 × **7** = 21

21 ÷ 3 = **7**

Use the fact triangle.

Find the unknown.

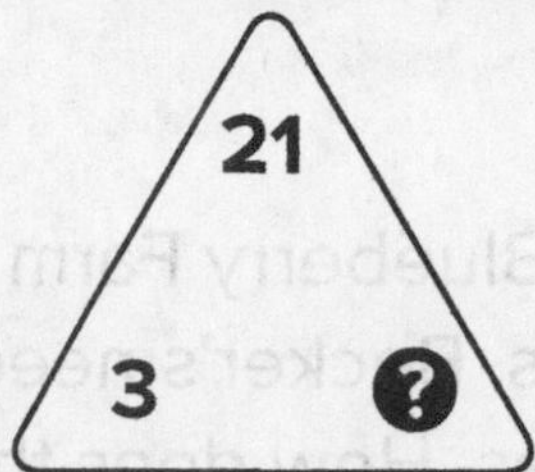

Divide by 6

30 ÷ 6 = **?**

Think: 6 × **?** = 30

30 ÷ 6 = **5**

6 × **5** = 30

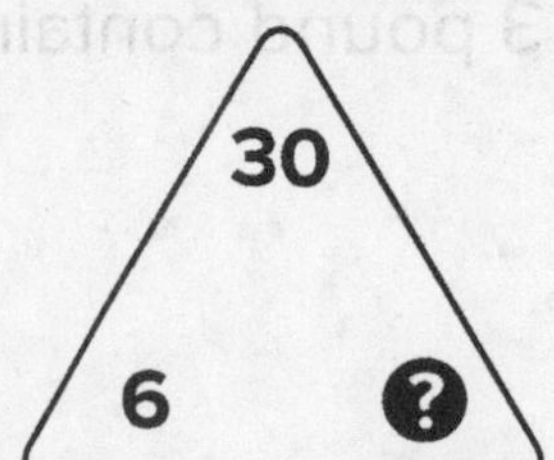

Complete the equation to make it true.

1. 6 ÷ 6 = ________

2. ________ = 24 ÷ 6

3. ________ = 36 ÷ 6

4. 48 ÷ 6 = ________

5. ________ = 15 ÷ 3

6. 12 ÷ 3 = ________

Find the unknown number in the fact triangle. Write the four related facts.

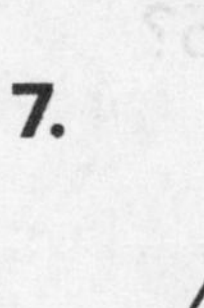

7.

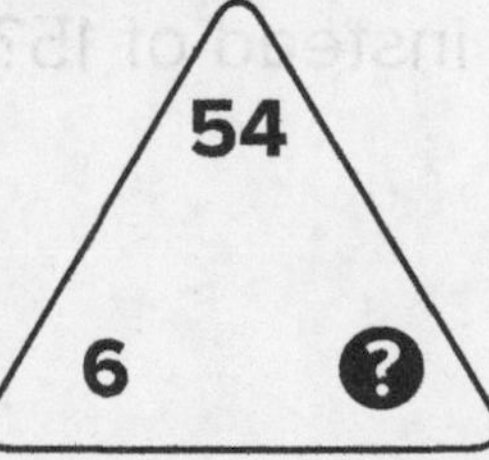

________________ ________________

________________ ________________

Divide by 3 and 6

Name ____________________

Solve. Explain your work.

1. April is packing peaches in baskets of 6. She needs baskets for 60 peaches. How does the number of baskets compare if she packs 12 in each basket instead of 6?

2. Becker's Blueberry Farm is shipping blueberries in 6 lb. containers. Becker's needs containers for 30 pounds of blueberries. How does the number of containers compare if they ship 3 pound containers instead of 6 pound containers?

3. Cedric and his brothers picked 24 pounds of oranges. They want to pack boxes of 6 pounds each. How does the number of boxes compare if they pack boxes of 3 pounds instead of 6 pounds?

4. A diner ordered 90 pears from the Baker Fruit Company. They usually pack 15 pears in each crate. They want to pack crates of 30 pears this time. How does the number of crates compare if they pack 30 pears in each crate instead of 15?

Divide by 4 and 8

Name ________________________________

Review

You can use the relationship between multiplication and division to divide by 4 and 8.

Divide by 4 $12 \div 4 = ?$ **Think:** $4 \times ? = 12$ $4 \times \mathbf{3} = 12$ $12 \div 4 = \mathbf{3}$	**Use the fact triangle to help you find a related multiplication fact.** 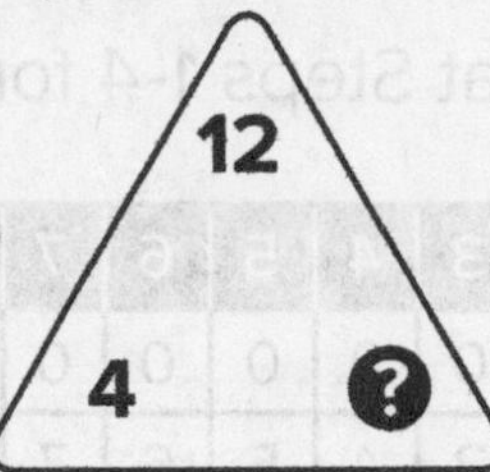
Divide by 8 $24 \div 8 = ?$ **Think:** $8 \times ? = 24$ $24 \div 8 = \mathbf{3}$ $8 \times \mathbf{3} = 24$	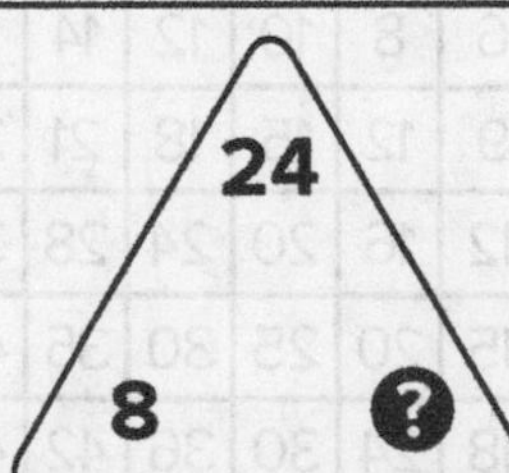

Complete the equation to make it true.

1. $56 \div 8 =$ ______ **2.** ______ $= 64 \div 8$

3. ______ $= 36 \div 4$ **4.** $8 \div 4 =$ ______

5. $32 \div 4 =$ ______ **6.** ______ $= 8 \div 8$

Find the unknown number in the fact triangle. Write the four related facts.

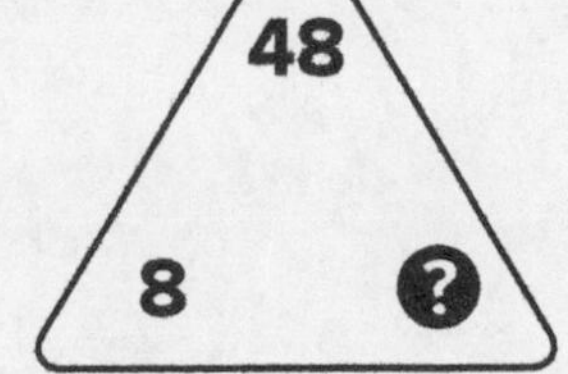

7. 

Divide by 4 and 8

Name ______________________________

Invent your own method or follow Steps 1–5 to write rules for when you can divide any number by 4 or by 8.

Step 1 Circle products of numbers multiplied by 4.

Step 2 Look for patterns.

Step 3 Think of related division facts.

Step 4 Write a rule to tell when a number can be divided by 4.

Step 5 Repeat Steps 1-4 for numbers multiplied by 8.

×	0	1	2	3	4	5	6	7	8	9	10
0	0	0	0	0	0	0	0	0	0	0	0
1	0	1	2	3	4	5	6	7	8	9	10
2	0	2	4	6	8	10	12	14	16	18	20
3	0	3	6	9	12	15	18	21	24	27	30
4	0	4	8	12	16	20	24	28	32	36	40
5	0	5	10	15	20	25	30	35	40	45	50
6	0	6	12	18	24	30	36	42	48	54	60
7	0	7	14	21	28	35	42	49	56	63	70
8	0	8	16	24	32	40	48	56	64	72	80
9	0	9	18	27	36	45	54	63	72	81	90
10	0	10	20	30	40	50	60	70	80	90	100

2. What you could tell someone to help them think of a strategy for recalling division facts for 4 and 8?

Divide by 9

Name ________________________________

Review

You can use related multiplication facts to divide by 9.

Divide by 9

$18 \div 9 = ?$

Think: $9 \times ? = 18$

$9 \times \mathbf{2} = 18$

$18 \div 9 = \mathbf{2}$

Use an array to help you find a related multiplication fact.

Another Way

$27 \div 9 = ?$

Think: $9 \times ? = 27$

$9 \times \mathbf{3} = 27$

$27 \div 9 = \mathbf{3}$

Use the fact triangle to help you find a related multiplication fact.

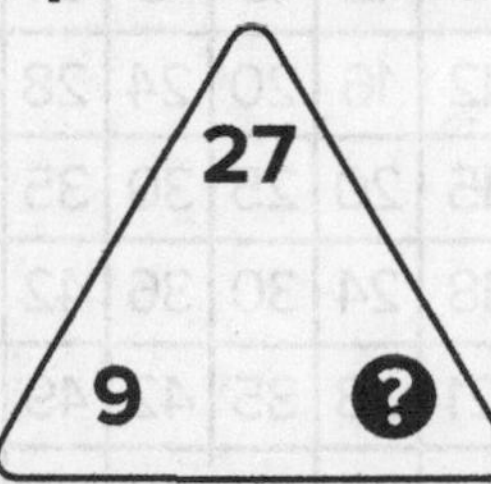

Complete the equation to make it true.

1. $45 \div 9 =$ ________

2. ________ $= 36 \div 9$

3. ________ $= 9 \div 9$

4. $54 \div 9 =$ ________

5. $81 \div 9 =$ ________

6. ________ $= 63 \div 9$

Write two related division equations for the multiplication fact.

7. $9 \times 6 = 54$

8. $9 \times 8 = 72$

Lesson 9-7 • Extend Thinking

Divide by 9

Name ______________________________

Use your own way or follow Steps 1-4 to write a rule that tells when you can divide a number by 9.

Step 1 Circle products of numbers multiplied by 9.

Step 2 Look for patterns in the products.

Step 3 Think of division facts related to the multiplication facts.

Step 4 Write a rule to tell when a number can be divided by 9.

×	0	1	2	3	4	5	6	7	8	9	10
0	0	0	0	0	0	0	0	0	0	0	0
1	0	1	2	3	4	5	6	7	8	9	10
2	0	2	4	6	8	10	12	14	16	18	20
3	0	3	6	9	12	15	18	21	24	27	30
4	0	4	8	12	16	20	24	28	32	36	40
5	0	5	10	15	20	25	30	35	40	45	50
6	0	6	12	18	24	30	36	42	48	54	60
7	0	7	14	21	28	35	42	49	56	63	70
8	0	8	16	24	32	40	48	56	64	72	80
9	0	9	18	27	36	45	54	63	72	81	90
10	0	10	20	30	40	50	60	70	80	90	100

Test your rule for each number. Justify your answer.

1. 99

2. 108

3. 117

4. 126

Lesson **9-8 • Reinforce Understanding**

Divide by 7

Name ______________________________

Review

You can use the relationship between multiplication and division to divide by 7. Use the multiplication fact table to help you find the unknown value.

Example

Divide: 21 ÷ 7 = **?**

Think: 7 × **?** = 21

7 × **3** = 21

21 ÷ 7 = **3**

×	0	1	2	3	4	5	6	7	8	9	10
0	0	0	0	0	0	0	0	0	0	0	0
1	0	1	2	3	4	5	6	7	8	9	10
2	0	2	4	6	8	10	12	14	16	18	20
3	0	3	6	9	12	15	18	21	24	27	30
4	0	4	8	12	16	20	24	28	32	36	40
5	0	5	10	15	20	25	30	35	40	45	50
6	0	6	12	18	24	30	36	42	48	54	60
7	0	7	14	21	28	35	42	49	56	63	70
8	0	8	16	24	32	40	48	56	64	72	80
9	0	9	18	27	36	45	54	63	72	81	90
10	0	10	20	30	40	50	60	70	80	90	100

Use the multiplication fact table to write a related multiplication fact and find the unknown value.

1. 63 ÷ 7 = ______

7 × ______ = ______

2. 28 ÷ 7 = ______

7 × ______ = ______

3. 14 ÷ 7 = ______

7 × ______ = ______

4. 49 ÷ 7 = ______

7 × ______ = ______

Complete the equation to make it true.

5. 35 ÷ 7 = ______

6. ______ ÷ 7 = ______

7. ______ = 56 ÷ 7

8. ______ = 63 ÷ 7

9. ______ = 42 ÷ 7

10. 70 ÷ 7 = ______

Divide by 7

Name ______________________________

Solve. Explain your work.

1. Kelly is buying craft sticks for summer camp activities. The store sells boxes of 8 packages for $7 each. She spends $56. How many boxes of craft sticks did Kelly buy?

2. Kelly is also buying bags of feathers for camp crafts. Three bags cost $7. How many bags of feathers can she buy with $21?

3. Mr. Rivera is helping Kelly buy supplies for camp. They are buying colorful rubber balls. A can has 7 rubber balls and costs $10. They buy 70 balls. How much do they spend on rubber balls?

4. Mr. Rivera is also buying jump ropes. He can buy a pack of 7 jump ropes for $10. He buys 63 jump ropes. How much does he spend?

5. Kelly and Mr. Rivera are buying juice boxes for the campers. One package has 7 juice boxes and costs $6. They buy 70 juice boxes. How much do they spend?

Multiply and Divide Fluently within 100

Name ________________________________

Review

Any multiplication or division strategy can be used to fluently multiply or divide. You can decompose a factor to find a product.

Decompose a factor to find $9 \times 7 = ?$

Break 7 into $2 + 5$

Multiply each part by 9. $9 \times 2 = 18$ $9 \times 5 = 45$

Add the products $18 + 45 = 63$

$9 \times 7 = 63$

Complete the equation to make it true.

1. $6 \times$ ______ $= 24$
2. ______ $= 40 \div 10$
3. $8 = 48 \div$ ______
4. ______ $= 7 \times 7$
5. $6 =$ ______ $\div 2$
6. ______ $\times 9 = 27$

Decompose 7 × 8 to find the product. Circle all the correct answers.

7. $7 \times 4 + 7 \times 4$ $4 \times 7 + 2 \times 4$

$7 \times 5 + 3 \times 7$ $4 \times 2 + 4 \times 7$

Lesson **9-9 • Extend Thinking**

Multiply and Divide Fluently within 100

Name ______________________________

Solve.

1. Kiara uses 18 cups of flour to make 3 different kinds of muffins for a bake sale. She uses equal amounts of flour for each kind of muffin. How many cups of flour does she use to make each kind of muffin?

2. Liam uses 16 cups of popcorn to make 4 kinds of popcorn balls for the bake sale. He uses equal amounts of popcorn for each kind of popcorn ball. How many cups of popcorn does he use to make each kind of popcorn balls?

3. Ari uses 24 cups of oatmeal to make 8 kinds of snack bars for the bake sale. He uses equal amounts of oats for each kind of bar. How many cups of oatmeal does he use to make each kind?

4. Skylar makes zucchini bread for the bake sale. Skylar uses a total of 28 cups of shredded zucchini. She uses an equal amount of shredded zucchini the in batter for 7 loaves of zucchini bread. How many cups of shredded zucchini does she use in each loaf?

Patterns with Multiples of 10

Name ____________________

Review

You can think about how many tens the product will have.	You can decompose (break apart) the multiple of 10 to find the product.
Think: $6 \times 30 = ?$	**Think:** $6 \times 30 = ?$
$6 \times \mathbf{30} = 6 \times \mathbf{3\ tens}$	$6 \times \mathbf{30} = 6 \times \mathbf{3} \times \mathbf{10}$
$= 18$ tens	$= 18 \times 10$
$= 180$	$= 180$

How many tens will the product have? What is the product?

1. $3 \times 20 = ?$

= ________ tens

= ________

2. $5 \times 40 = ?$

= ________ tens

= ________

3. $4 \times 30 = ?$

= ________ tens

= ________

4. $6 \times 70 = ?$

= ________ tens

= ________

How can you decompose the multiple of 10 to find the product? Fill in the blanks to show how.

5. $5 \times 70 = 5 \times$ ________ $\times 10$

= ________ $\times 10$

= ________

6. $4 \times 80 = 4 \times$ ________ $\times 10$

= ________ $\times 10$

= ________

Lesson **10-1 • Extend Thinking**

Patterns with Multiples of 10

Name ____

Solve two riddles, then write two riddles.

1. I am a multiplication sentence that uses a multiple of 10. I have a product of 240. What multiplication sentence am I?

______ × ______ = ______

______ × ______ = ______

2. I am a multiplication sentence that uses a multiple of 10. I have a product of 320. What multiplication sentence am I?

______ × ______ = ______

______ × ______ = ______

3. ____

____

______ × ______ = ______

______ × ______ = ______

4. ____

____

______ × ______ = ______

______ × ______ = ______

More Multiplication Patterns

Name ______________________________

Review

You can use a multiplication table to find patterns with factors and products.

Example

Products of 6 are double the products of 3.

Example

Products of 7 have an odd and even pattern.

×	0	1	2	3	4	5	6	7	8	9	10
0	0	0	0	0	0	0	0	0	0	0	0
1	0	1	2	3	4	5	6	7	8	9	10
2	0	2	4	6	8	10	12	14	16	18	20
3	0	3	6	9	12	15	18	21	24	27	30
4	0	4	8	12	16	20	24	28	32	36	40
5	0	5	10	15	20	25	30	35	40	45	50
6	0	6	12	18	24	30	36	42	48	54	60
7	0	7	14	21	28	35	42	49	56	63	70
8	0	8	16	24	32	40	48	56	64	72	80
9	0	9	18	27	36	45	54	63	72	81	90
10	0	10	20	30	40	50	60	70	80	90	100

How can you describe the pattern? Use the multiplication table.

1. Products of 4

2. Products of 5

3. Products of 3

4. Products of 10

Which facts have an even product? Circle the facts.

5. 4 × 3   4 × 6   1 × 3   2 × 3

6. 5 × 9   5 × 4   8 × 7   7 × 7

Which facts have an odd product? Circle the facts.

7. 2 × 5   3 × 5   4 × 7   5 × 7

8. 7 × 9   9 × 9   8 × 9   8 × 8

More Multiplication Patterns

Name ______________________________

Use different colors to show patterns for factors and products.

×	0	1	2	3	4	5	6	7	8	9	10
0	0	0	0	0	0	0	0	0	0	0	0
1	0	1	2	3	4	5	6	7	8	9	10
2	0	2	4	6	8	10	12	14	16	18	20
3	0	3	6	9	12	15	18	21	24	27	30
4	0	4	8	12	16	20	24	28	32	36	40
5	0	5	10	15	20	25	30	35	40	45	50
6	0	6	12	18	24	30	36	42	48	54	60
7	0	7	14	21	28	35	42	49	56	63	70
8	0	8	16	24	32	40	48	56	64	72	80
9	0	9	18	27	36	45	54	63	72	81	90
10	0	10	20	30	40	50	60	70	80	90	100

Fill in the blank to make the statement true. Write "always", "sometimes", or "never".

1. Products of 10 are ____________ products of 5.

2. Products of 3 are ____________ products of 5.

3. Products of 7 are ____________ even.

4. Products of 6 are ____________ double the products of 3.

5. Products of 9 are ____________ double the products of 3.

6. Products of 9 are ____________ products of 3.

7. Products of 3 are ____________ triple the products of 1.

8. Products of 2 are ____________ double the products of 1.

9. Odd number products are ____________ odd.

10. Even number products are ____________ odd.

Understand the Associative Property

Name ______________________________

Review

You can group factors in different ways to multiply. It does not change the product. This is a property (or rule) of multiplication.

Example 1 $4 \times 2 \times 3 = ?$	**Example 2** $4 \times 2 \times 3 = ?$
Multiply the **first two factors.**	Multiply the **second two factors.**
Think $\mathbf{4 \times 2} \times 3 = ?$	**Think** $4 \times \mathbf{2 \times 3} = ?$
$\mathbf{8} \times 3 = 24$	$4 \times \mathbf{6} = 24$
So, $4 \times 2 \times 3 = 24$.	

How can you use the Associative Property to solve the equation two ways? Fill in the blanks to show how.

1. $3 \times 2 \times 2 = ?$ $\quad 3 \times 2 \times 2 = ?$

   $= 3 \times 2 \times$ ___ $\quad =$ ___ $\times 2 \times 2$

   $=$ ___ $\times$ ___ $\quad =$ ___ $\times$ ___

   $=$ ___ $\quad =$ ___

2. $4 \times 2 \times 5 = ?$ $\quad 4 \times 2 \times 5 = ?$

   $= 4 \times 2 \times$ ___ $\quad =$ ___ $\times 2 \times 5$

   $=$ ___ $\times$ ___ $\quad =$ ___ $\times$ ___

   $=$ ___ $\quad =$ ___

Lesson **10-3 • Extend Thinking**

Understand the Associative Property

Name ______________________________

How can you use the Associative Property to make multiplying easier? Solve each problem two ways. Is one way easier? Why?

1. Alex writes 7 pages for each of 2 stories 10 days a month. How many pages does Alex write in each month?

2. Cameron swam 2 laps around an Olympic pool 4 days a week. He did this for 10 weeks in a row. How many laps did he swim in all?

3. Ava practices piano for 2 hours a day 5 days a week. How many hours does she practice in 3 weeks?

4. Lucas plays soccer after school for 1 hour a day 5 days a week. How many hours does he play in 10 weeks?

Lesson **10-4 • Reinforce Understanding**

Two-Step Problems Using Multiplication and Division

Name ______________________________

Review

You can make a diagram for each part of a two-step problem and then write an equation for each diagram.

Problem Mel brings cantaloupe slices to baseball camp. She needs 2 slices each for 6 campers, but the slices are in bags of 4. How many bags of cantaloupe slices does Mel need?

Diagram	Equation	Answer
C C C C C C (each branching to ss)	$2 \times 6 = s$ $s = 12$	She needs 12 slices of cantaloupe.
ssss ssss ssss (each joining to B)	$12 \div 4 = B$ $B = 3$	She needs 3 bags of cantaloupe slices.

How can you use equations with a letter for the unknown to solve the problem?

1. Ty has a collector's album of sports trading cards with 6 pages. Each page has 4 cards. He wants to rearrange the cards to place all of them on 8 pages. If he puts an equal number on each page, how many cards will he put on each page?

2. Shira brings home 48 trading cards from the trading card show and divides the cards into 8 equal groups. She keeps only 1 group for herself and then gives half of her group to her sister. How many cards does Shira have left for herself?

Lesson **10-4 • Extend Thinking**

Two-Step Problems Using Multiplication and Division

Name ______________________________

How can you solve the two-step problem? Write your answer and explain how you found it.

1. Mr. Lipscomb buys 6 packages of tee shirts. He gives 4 tee shirts to each student. If 9 students get tee shirts, how many tee shirts came in a package?

2. Mr. Lipscomb buys 10 packages of fabric paint pens. He gives 5 pens to each student. If 6 students get fabric paint pens, how many came in a package?

3. Mr. Lipscomb buys 10 packages of bibs and gives an equal number to each of his 8 grandbabies. There are 4 bibs in each package. How many bibs will each baby get?

4. Mr. Lipscomb needs 48 rolls of decorative ribbon for a project. The ribbon is sold in packages of 6. The order is shipped in 2 boxes. How many packages of ribbon are in each box?

Solve Two-Step Problems

Name ______________________________

Review

You can represent a two-step problem with two equations.

Problem Ana buys 1 pair of flip-flops and 4 towels. How much does she spend?

Towels $12
Flip-flops $10

Step 1: Write an equation for the cost of 4 towels.

$c =$ cost of towels

$c = 4 \times 12$

$c = 48$

Step 2: Write an equation for the total cost of the towels and the flip-flops

$T =$ total cost

$T = 48 + 10$

$T = 58$

Step 3: Write the final answer. Ana spends $58

How can you use 2 equations to solve the problem?

1. Maria sells muffins in packs of 6. If Sue buys 9 packs and 16 individual muffins, how many muffins does she buy in all?

2. Deion shares 28 grapes equally among 4 friends. Then he gives each friend 5 additional grapes. How many grapes does each friend receive?

3. After school, 24 children lined up to try an obstacle course. Then 4 children decided to play with a flying disc instead. The rest of the children made teams of 5 each. How many teams were there?

Solve Two-Step Problems

Name ____

Do you *agree* or *disagree* with the solution given? Circle your answer and explain your reasoning.

1. Adriana babysits for 2 hours on each of 4 days in a week. Then she babysits 4 hours on another day. She calculates the number of hours she has spent babysitting.

   $2 + 6 = h$ $h = 8$ agree disagree

   $8 \times 4 = t$ $t = 32$ She works 32 hours.

2. Lewis dog sits for 3 weekdays on each of 4 weeks in a month. He also dog sits all weekend one week in a month. He calculates the number of days he has spent dog sitting.

   $3 \times 4 = d$ $d = 12$ agree disagree

   $12 + 2 = t$ $t = 14$ He dog sits 14 hours a month.

3. Dale house sits for 5 days on each of 2 weeks in month. Then he house sits 2 weekends a month. He calculates the number of days he has spent house sitting.

   $2 + 5 = d$ $d = 7$ agree disagree

   $2 \times 5 = t$ $t = 10$ He house sits 17 days a month.

Explain the Reasonableness of a Solution

Name ______________________________

Review

You can use mental math and estimation to determine the reasonableness of an answer to a two-step problem.

Problem A bookcase has 8 shelves with 6 books on each shelf. Lin reads 22 of the books. He says that he has 12 books left to read. Determine if Lin's statement is reasonable.

Estimate	Use Mental Math
$b = 8 \times 6$ $b = 48$ **Think:** use $b = 50$	$48 - 22 = L$ **Think:** $50 - 20 = L$ $L = 30$ books left

Lin has 30 books left to read, so 12 is not reasonable.

How can you estimate to determine the reasonableness of an answer? Circle the reasonable answer.

1. At the bowling alley, Mia buys snacks for \$2 and 3 bowling passes for \$6 each. How much does Mia spend?

A. \$30 **B.** \$20 **C.** \$18 **D.** \$200

2. June shares 18 crackers evenly among 9 friends. Then she gives each friend 2 more crackers. How many crackers does each friend receive?

A. 4 crackers **B.** 40 crackers
C. 7 crackers **D.** 29 crackers

3. At the skating rink, Dana pays \$14 for admission and rents 2 pairs of roller skates for \$2 each. How much does she spend?

A. \$70 **B.** \$18 **C.** \$28 **D.** \$142

Lesson **10-6 • Extend Thinking**

Explain the Reasonableness of a Solution

Name ______________________________

Is the solution reasonable? Explain how you know.

1. Cindy has a box of 36 granola bars. She eats 4 and wants to give the rest to 5 friends to share equally. She estimates there are enough for each friend to get 6 granola bars.

2. Manny has 42 toy cars. He displays 15 and wants to give the rest to 3 friends to share equally. He estimates there are enough for each friend to get 8 toy cars.

3. Lee has a collection of 52 small stuffed animals. He keeps 9 and wants to give the rest to 8 friends to share equally. He estimates there are enough for each friend to get 5.

4. Yanni has a box of 30 action figures. He plays with 12 and wants to give the rest to 5 friends to share equally. He estimates there are enough for each friend to get 5.

Lesson **11-1 • Reinforce Understanding**

Understand Perimeter

Name ________________________________

Review

You can use different strategies to find the perimeter of a figure.

Hint:
The distance around a figure is called the perimeter.

Count the side lengths around the figure to find the perimeter.

— = 1 length

12 cm

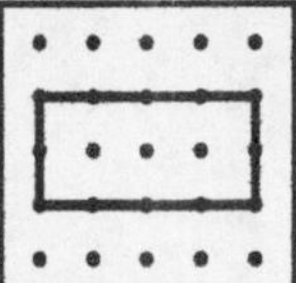

Add the side lengths.

$4 + 2 + 4 + 2 = 12$

The perimeter of the figure is 12 cm.

4 cm
2 cm
2 cm
4 cm

Use an equation to find the perimeter. Choose the correct equation for the figure.

1.

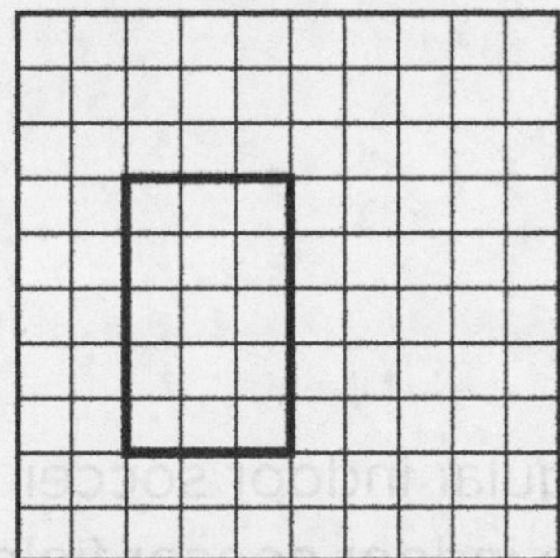

A. 5×3

B. $5 + 3$

C. $5 + 3 + 5 + 3$

D. $5 \times 3 \times 5 \times 3$

2.

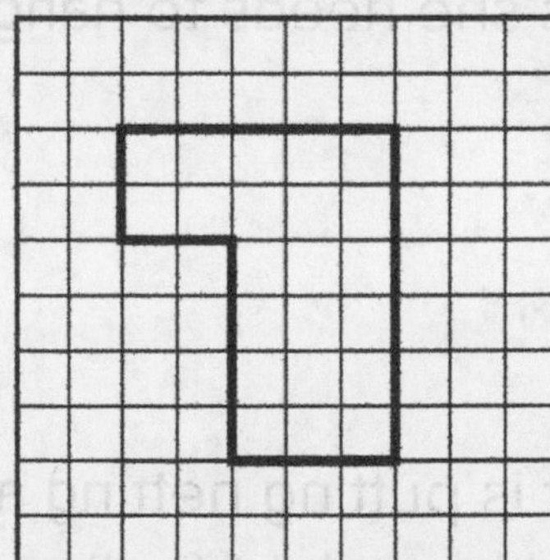

A. $4 + 7 + 5 + 2 + 3 + 3$

B. $3 + 6 + 5 + 3 + 2 + 4$

C. $6 + 3 + 4 + 2 + 2 + 5$

D. $7 + 5 + 2 + 3 + 3 + 6$

Understand Perimeter

Name ______________________________

Solve.

1. Oliver is building a fence to go around a rectangular playground. How might finding perimeter of the playground help him complete the job?

2. Lila is laying a wooden frame around a square picnic area. How might finding perimeter of the picnic area help her complete the frame?

3. Sarika is building a fence around a basketball court. She needs fence posts and fencing. How might finding perimeter of the basketball court help Sarika figure out how many fence posts she needs to hang the fence on?

4. Harjit is putting netting around a rectangular indoor soccer field. How might finding perimeter of the indoor soccer field help him complete the project?

Determine the Perimeter of Figures

Name ____________________

Review

You can use strategies to determine the perimeter of a figure.

Add the side lengths around the rectangle to find the perimeter.

4 + 2 + 4 + 2 = 12

The perimeter of the rectangle is 12 yards.

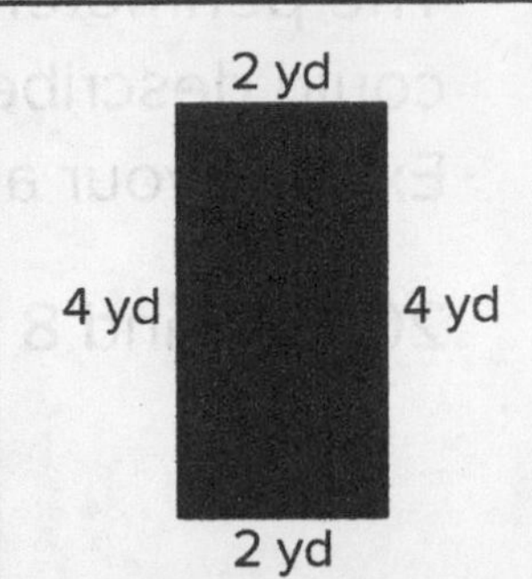

Add the side lengths or multiply the side length by 4.

5 + 5 + 5 + 5 = 20

4 × 5 = 20

The perimeter of the square is 20 yards.

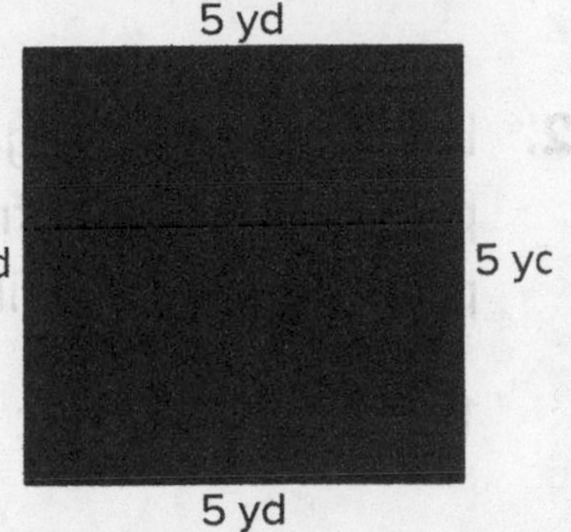

Complete the equation. Find the perimeter of the figure.

1.

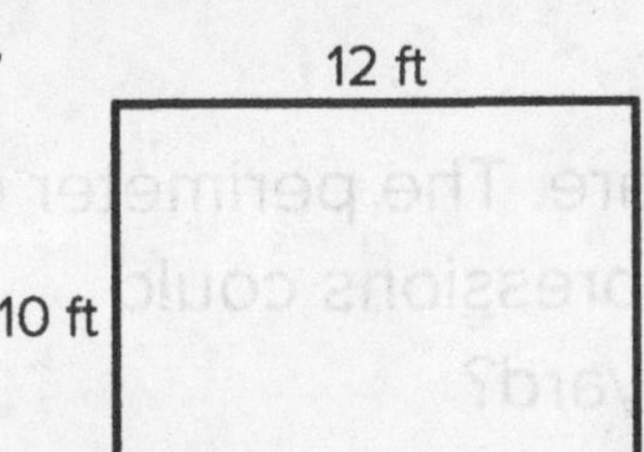

_____ + _____ + _____ + _____

Perimeter = _____ feet

Find the perimeter of the figure. Label with units.

2. __________

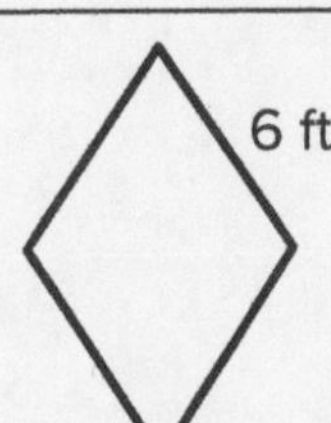

3. __________

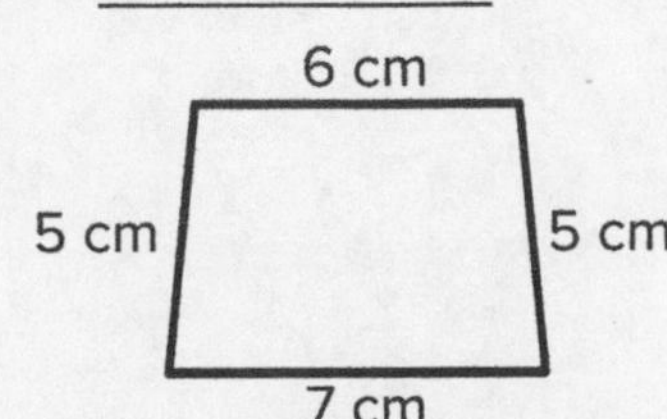

4. __________

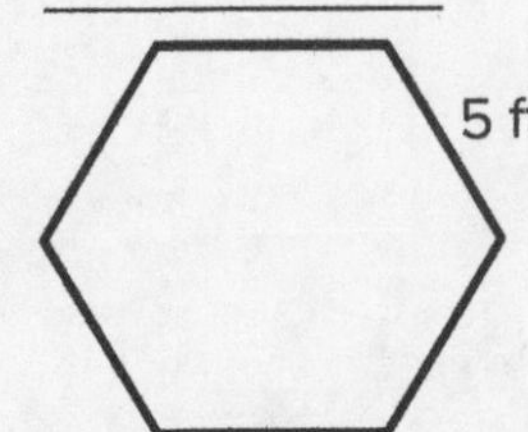

Determine the Perimeter of Figures

Name ______________________________

Solve. Explain your answer.

1. Param's backyard is in the shape of a rectangle. The perimeter of his yard is 60 feet. Which of these pairs could describe the two side lengths of Param's backyard? Explain your answer.

20 feet and 8 feet   30 feet and 30 feet   21 feet and 9 feet

2. Lian's front yard garden is in the shape of a rectangle. The perimeter of her front yard garden is 30 feet. Which of these pairs could describe the two side lengths of Lian's garden?

12 feet and 4 feet   10 feet and 10 feet   6 feet and 9 feet

3. Aliyah's side yard is in the shape of a square. The perimeter of her side yard is 12 feet. Which of these expressions could describe the side lengths of Aliyah's side yard?

$4 + 4 + 4 + 4$ $3 + 3 + 3 + 3$ 4×4

Lesson **11-3 • Reinforce Understanding**

Determine an Unknown Side Length

Name ______________________________

Review

You can determine an unknown side length by subtracting the sum of the known side lengths from the perimeter.

The perimeter of the rectangle is 50 ft.

Step 1 Use the perimeter and the known sides to write an equation. 50 = 15 + 15 + ? + ?

Step 2 Simplify to 50 = 30 + ? + ?.

Step 3 Subtract the known side lengths from the perimeter.

50 − 30 = ? + ?

20 = **10 + 10**

The unknown side is 10 ft.

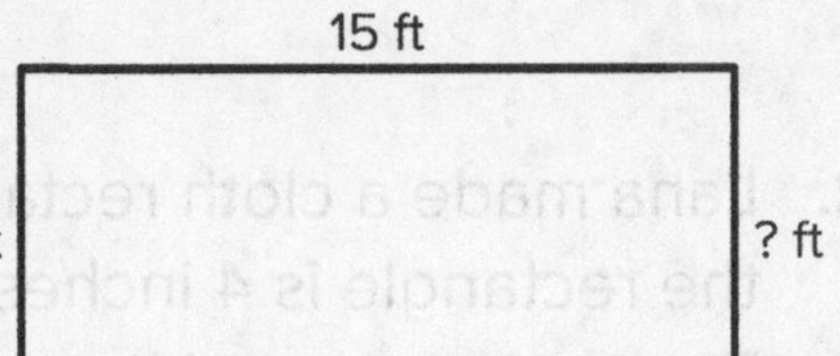

Find the unknown side length(s) of the figure.

1. The perimeter is 42 yards.

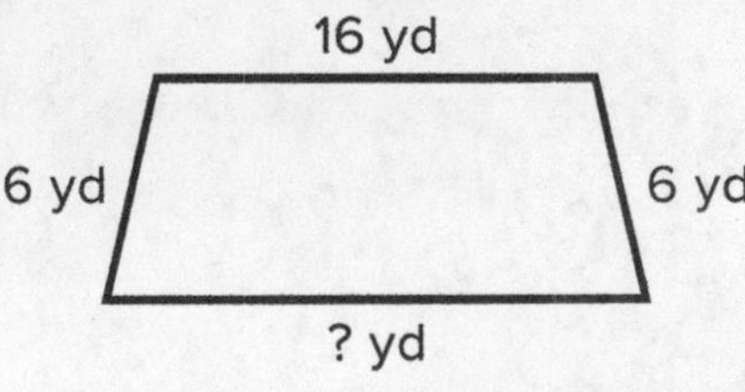

42 = 6 + 16 + __________ + ?

42 = __________ + ?

42 − __________ = ?

__________ = ?

2. The perimeter is 26 feet.

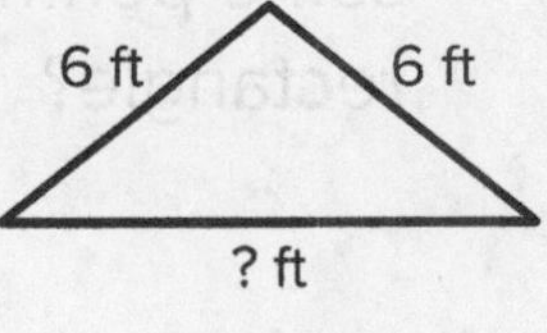

26 = __________ + ? + __________

26 = __________ + ?

26 − __________ = ?

__________ = ?

Determine an Unknown Side Length

Name ______________________________

Solve. Explain your answers.

1. Brita cut out a paper rectangle and a paper square. One side of the rectangle is 4 inches. Another side is twice as long. The rectangle and the square have the same perimeter. What is the side length of the square?

2. Lana made a cloth rectangle and a cloth square. One side of the rectangle is 4 inches. Another side is three times as long. The rectangle and the square have the same perimeter. What is the side length of the square?

3. David painted a rectangle and a square. One side of the square is 6 inches. The rectangle and the square have the same perimeter. What are the opposite side lengths of the rectangle?

Solve Problems Involving Area and Perimeter

Name ______________________________

Review

You can write addition equations to find perimeters and areas.

Hint:
Figures with the same area can have different perimeters.
Figures with the same perimeter can have different areas.

Example 1	Example 2	Example 3
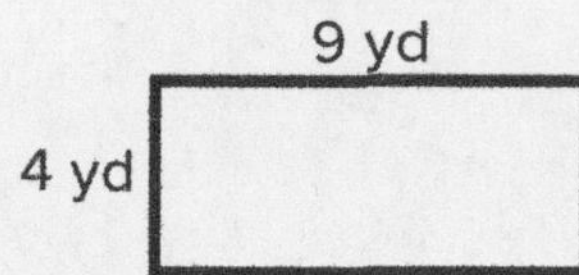	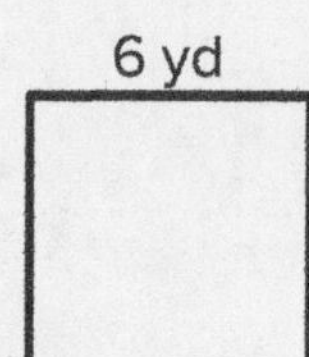	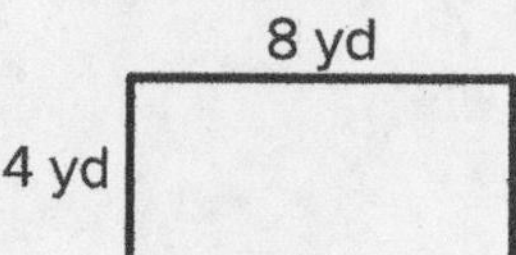
Perimeter $4 + 9 + 4 + 9 =$ 26 yds	**Perimeter** $6 + 6 + 6 + 6 =$ 24 yds	**Perimeter** $4 + 8 + 4 + 8 =$ 24 yds
Area $4 \times 9 = 36$ sq. yds	**Area** $6 \times 6 = 36$ sq. yds	**Area** $4 \times 8 = 32$ sq. yds

Find the perimeter and area of the figure. Include the unit.

1. 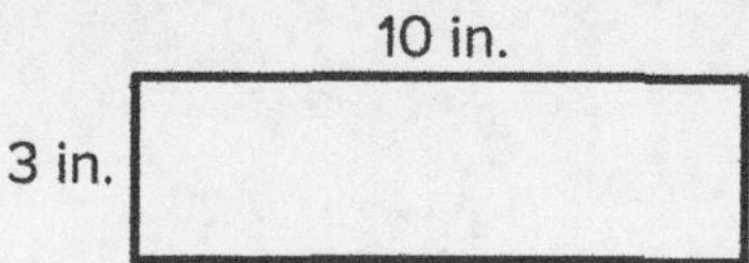

perimeter = ____________

area = ____________

2. 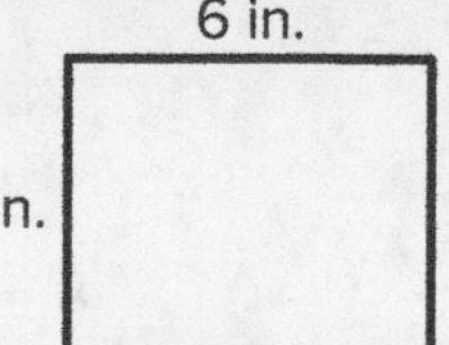

perimeter = ____________

area = ____________

3. 

perimeter = ____________

area = ____________

4. 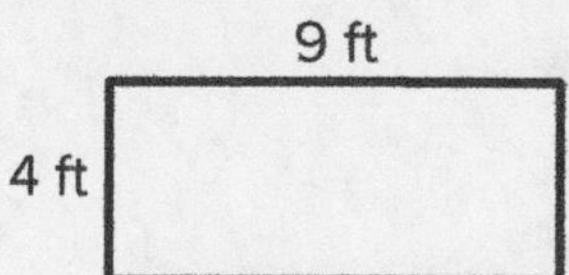

perimeter = ____________

area = ____________

Solve Problems Involving Area and Perimeter

Name ____________________

Draw and label two or more figures with the same area but different perimeters. Be sure to include the units.

Draw and label three figures with the same perimeter but different areas. Be sure to include units.

Solve Problems Involving Measurement

Name ______________________________

Review

You can solve word problems by representing the problem with an equation and using strategies to solve the equation.

Example

Mia connects 3 floorboards. The total length of the 3 floorboards is 24 inches. How long is each floorboard?

Step 1 Use a model to represent the problem.

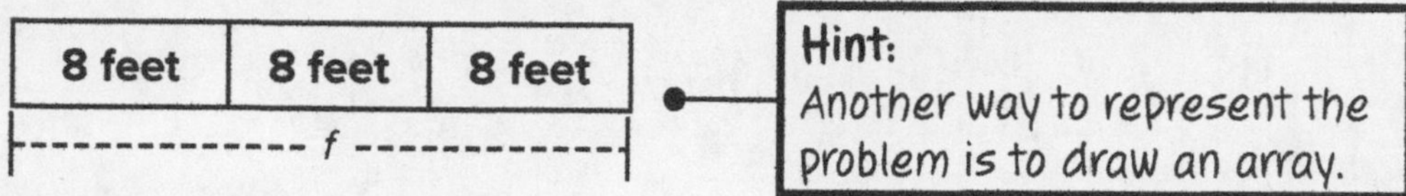

Step 2 Write an equation to represent the problem.

$3 \times 8 = f$

$3 \times 8 = 24$

Draw a model and write an equation to represent and solve the problem.

1. There are 4 plastic toy bins lined up against Jared's bedroom wall. Each bin is 2 feet long. What is the total length of the 4 toy bins?

2. A ribbon is 54 inches long. Stella plans to cut the ribbon into 9 equal pieces. How long is each piece?

3. Amber has 6 equal pieces of yarn. The total length of the 6 pieces of yarn is 48 inches. How long is each piece?

Solve Problems Involving Measurement

Name ______________________________

Show two different models to represent the problem. Write an equation to represent each model. Explain why each model results in the same solution.

1. A spool holds 72 inches of craft sculpture wire. Selene plans to divide it into 8 equal pieces. How long will each piece be?

2. Matt has 8 rolls of fabric yarn. Each roll has 12 yards of fabric yarn. What is the total length of the 8 rolls of fabric yarn?

Measure Liquid Volume

Name ______________________________

Review

You can measure **liquid volume** in liters (L) and milliliters (mL).

It takes 1000 mL to a make 1 L.

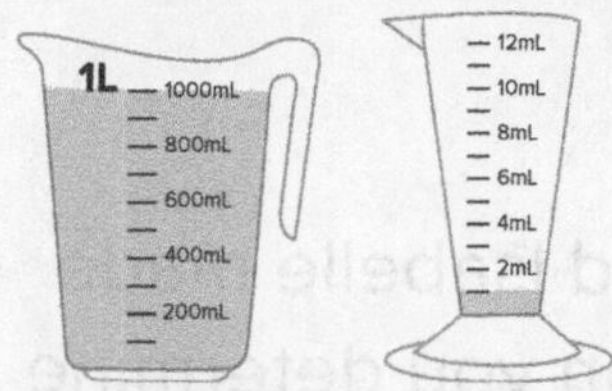

HINT
The amount of liquid in a container is **liquid volume**.

Write the liquid volume in the measuring cup.

1. _____ milliliters **2.** _____ milliliters **3.** _____ liters

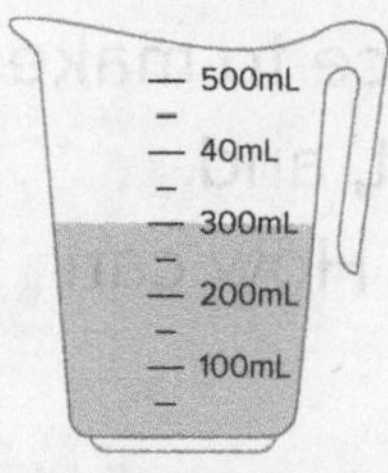

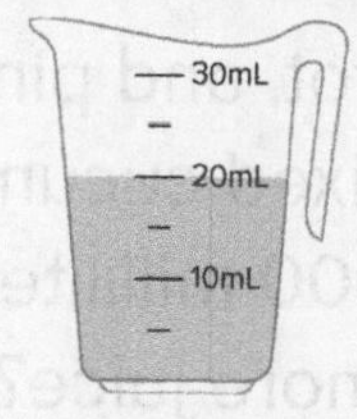

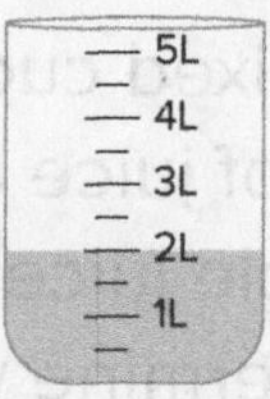

Circle the measuring cup that shows the given liquid volume.

4. 450 mL

A. 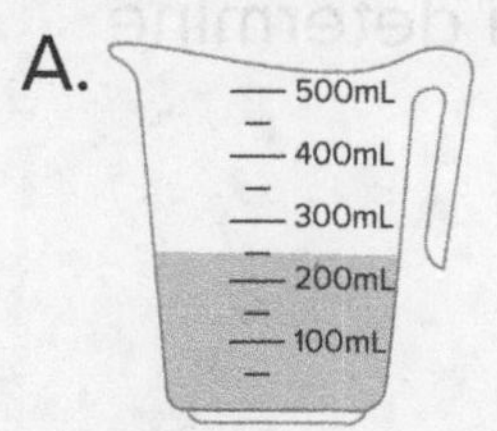

B. 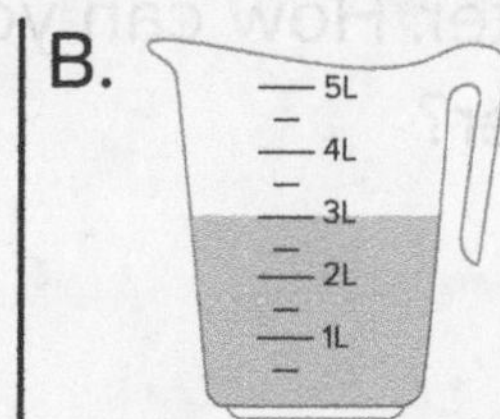

C. 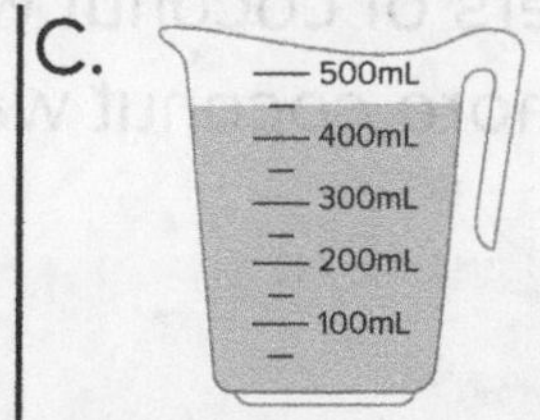

D. 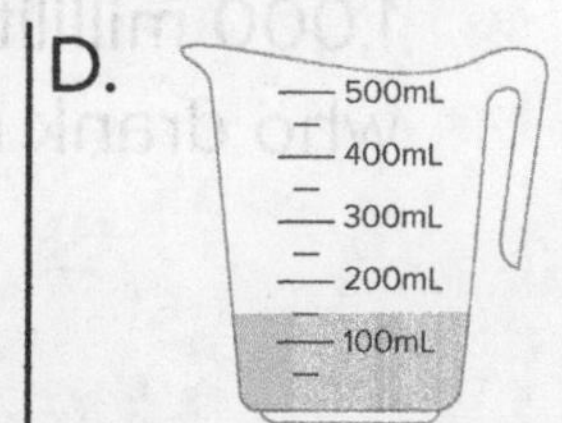

5. 3 liters

A. 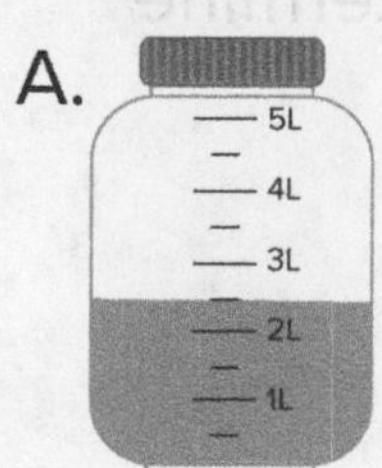

B. 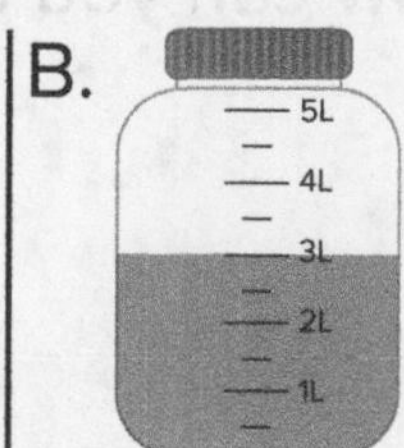

C. 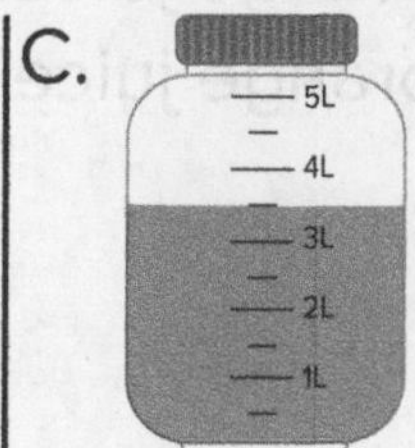

D. 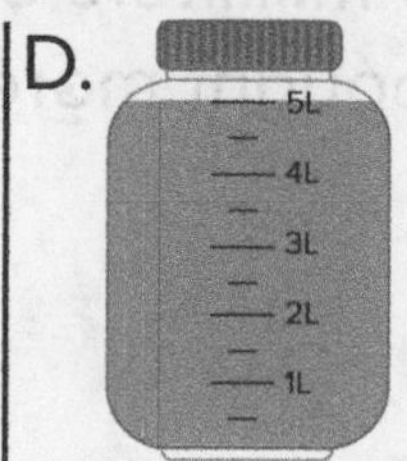

Measure Liquid Volume

Name ______________________________

Solve.

1. Aniq drank 500 milliliters of barley water and Alexis drank 1 liter of barley water. How can you determine who drank more barley water?

2. Ayden made 4 liters of lemonade and Isabelle made 1,000 milliliters of lemonade. How can you determine who made more lemonade?

3. Iyana mixed cucumber, carrot, and pineapple juice to make 2 liters of juice and Zack mixed cucumber, carrot, and pineapple juice to make 1,000 milliliters of juice. How can you determine who made more juice?

4. Jasmine drank 3 liters of coconut water and Ethan drank 1,000 milliliters of coconut water. How can you determine who drank more coconut water?

5. Wei bought 5 liters of orange juice and Jasmine bought 5,000 milliliters of orange juice. How can you determine who bought more orange juice?

Estimate and Solve Problems with Liquid Volume

Name ____

Review

You can think of containers to help you estimate units to use to measure liquid volumes.

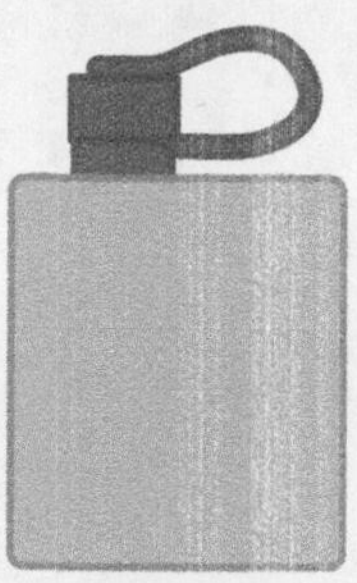

2 Liter
gas can

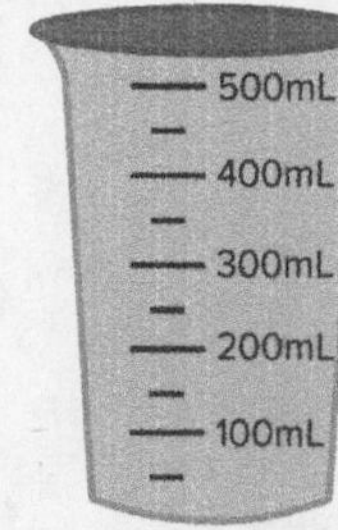

500 ml
cup

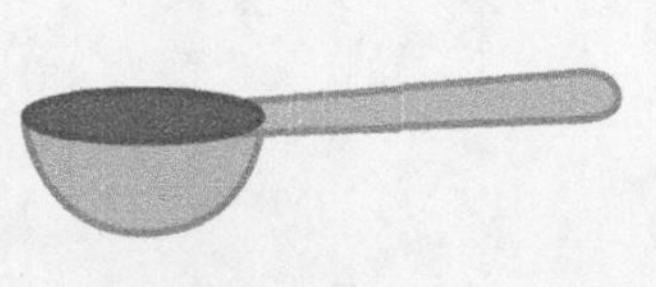

1 mL
spoon

Which is the better estimate for the liquid volume of the object?

1. 

40 liters

40 milliliters

4 liters

2. 

5 liters

500 milliliters

50 liters

3. 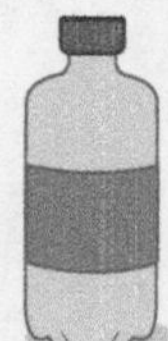

2 liters

200 milliliters

20 milliliters

Estimate and Solve Problems with Liquid Volume

Name ______________________________

Write three word problems that involve liquid volume. Solve. Write an equation to show your work.

1.

_____ × ________ = ________

2.

_____ × ________ = ________

3.

_____ × ________ = ________

Lesson **12-3 • Reinforce Understanding**

Measure Mass

Name ______________________

Review

You can use a balance scale to measure mass by adding weights until the scale is level.

HINT
Mass is the amount of matter in an object.

Think: Mass is measured in grams (g) and kilograms (kg).

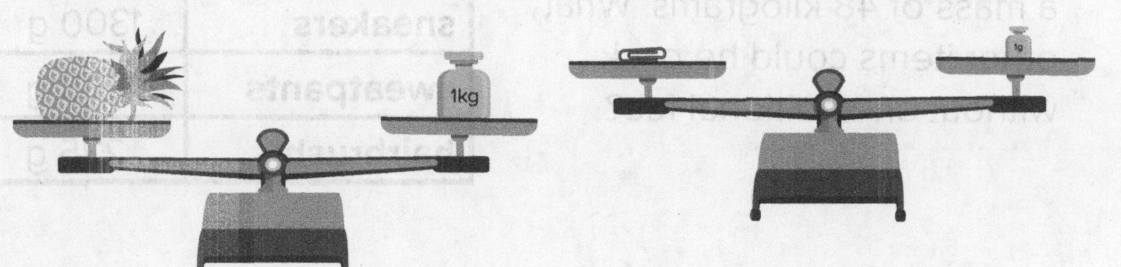

One kilogram (kg) is equal to 1,000 grams (g).

The weights show the mass of a fruit or vegetable. Write the mass in grams.

1. one small apple

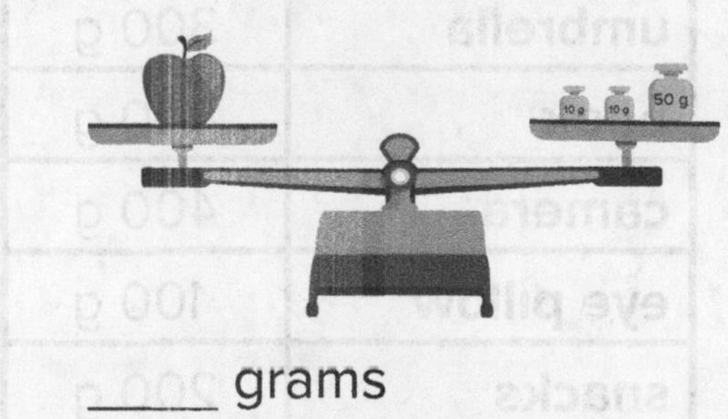

____ grams

2. bunch of grapes

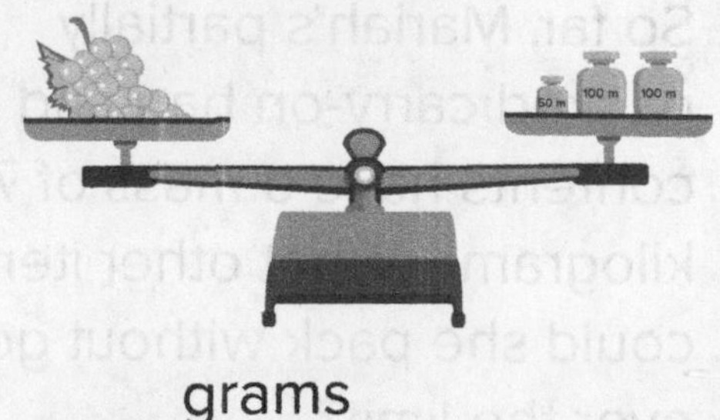

____ grams

3. bag of onions

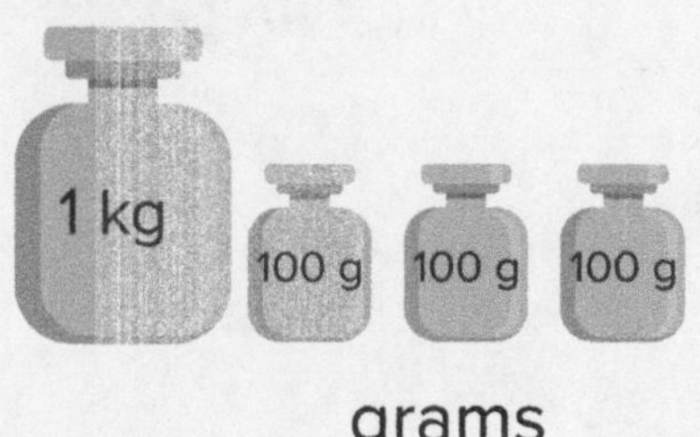

________ grams

4. four bananas

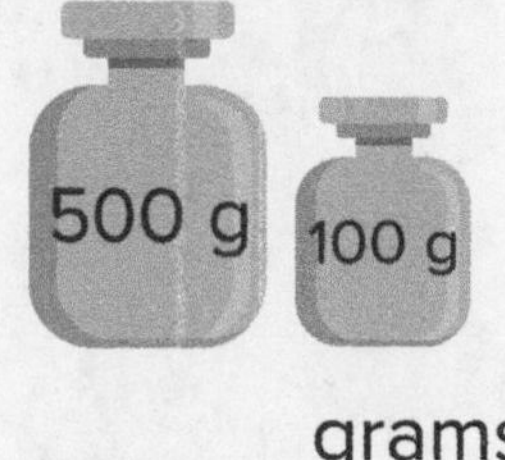

________ grams

Lesson **12-3 • Extend Thinking**

Measure Mass

Name ______________________________

Solve.

1. To travel on a train, a passenger can carry luggage with a mass of 50 kilograms for free. Aarav's packed luggage and its contents have a mass of 48 kilograms. What other items could he pack without an additional fee?

Item	Mass
bathing suit	150 g
pajamas	475 g
jeans	800 g
sneakers	1300 g
sweatpants	600 g
hairbrush	225 g

2. To travel on an airplane, a carry-on bag can have a mass of no more than 8 kilograms So far, Mariah's partially packed carry-on bag and its contents have a mass of 7 kilograms. What other items could she pack without going over the limit?

Item	Mass
blanket	150 g
umbrella	300 g
book	800 g
camera	400 g
eye pillow	100 g
snacks	200 g

Estimate and Solve Problems with Mass

Name ______________________________

Review

You can use everyday objects to help you estimate mass.

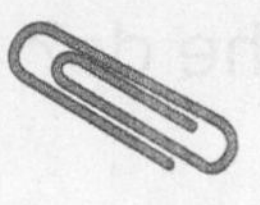

1 g

5 g

100 g

500 g

1 kg

Write the best estimate of the mass.

1. Which of the 3 weights is the best estimate for a can of beans?

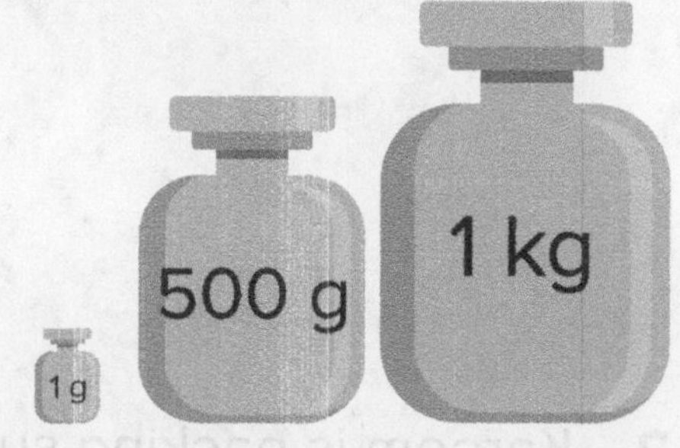

2. Which of the 3 weights is the best estimate for a box of cereal?

3. Which is the best estimate for a melon?

4. Which is the best estimate for a penny?

5. Which is the best estimate for a box of butter? Explain.

Lesson 12-4 • Extend Thinking

Estimate and Solve Problems with Mass

Name ______________________________

Solve

1. Shelia is packing a hiking bag for a family camping trip. Her parents don't want her bag to carry more than 12 kilograms. It weighs 14 kilograms. What should she do? Explain your reasoning.

- Remove a pair of socks?
- Remove a pair of field binoculars?
- Add a pair of sneakers?

2. Kareem is packing supplies for a day long bike ride. He wants ride with a cooler that weighs about 8 kilograms. The cooler weighs 6 kilograms now. What should he do? Explain your reasoning.

- Add extra water or juice bottles?
- Remove a notepad?
- Add a tire pump?

Lesson **12-5 • Reinforce Understanding**

Tell Time to the Nearest Minute

Name ______________________________

Review

You can count tick marks on an analog clock from a given number on the clock to tell time to the nearest minute.

analog **digital clock**

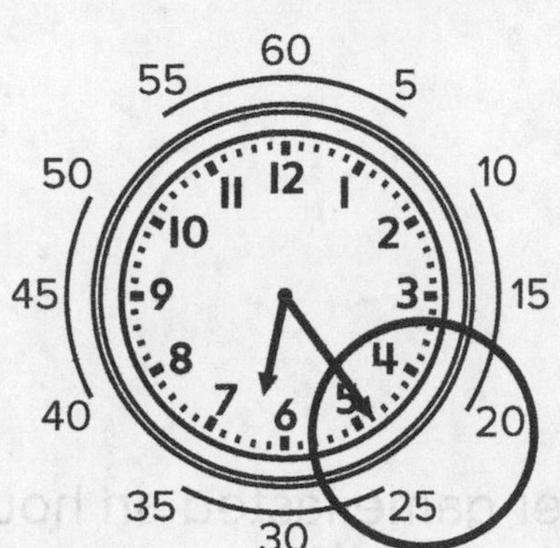

HINT
Each tick mark between the numbers on an analog clock represent 1 minute.

So, the minute hand on this analog clock shows that the time is 6:24. Read *6:24, 24 minutes past 6*, or *36 minutes before 7.*

Write the time shown on each clock.

1. 7: ____

2. 5: ____

3. ____ : ____

Draw the hands on the clock to show the given time.

4. 5:38

5. 2:12

6. 12:24

Lesson **12-5 • Extend Thinking**

Tell Time to the Nearest Minute

Name ______________________________

Solve. Draw hands on the clocks to show your answer.

1. Sammy went to play soccer at 3:27. He went home an hour later. What time did he go to soccer and when did he go home?

He went home at _____:_____

2. Kat went to play softball at 1:45. Her game lasted an hour and she went home 10 minutes later. What time did Kat's game end and when did she go home?

She went home at _____:_____

3. Yara went to play basketball at 12:32. Her sister joined her 2 hours later. What time did her sister join her?

Her sister joined her at _____:_____

Solve Problems Involving Time

Name ______________________________

Review

You can solve problems involving time intervals using an analog clock.

Example Jaquanda's haircut starts and ends at the times shown on the clocks.

Start

End

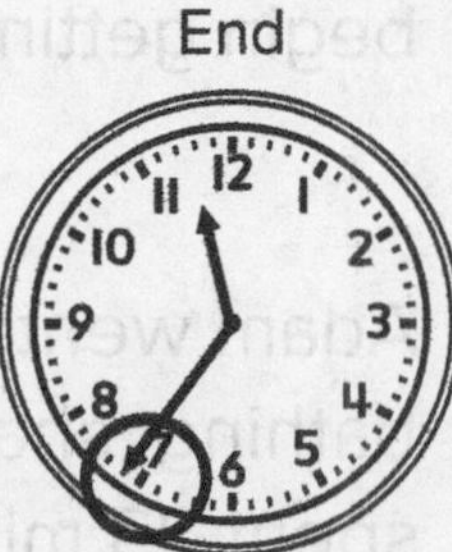

How long does her haircut take?

5 minute hops from 11:05 to 11:35.

1 minute hop from 11:35 to 11:36.

So, Jaquanda's haircut takes 31 minutes.

Find the time intervals. Use the clocks to answer questions 1–2.

1. What is the start time?

_____ : _____ p.m.

What is the end time?

_____ : _____ p.m.

2. What is the start time?

_____ : _____ p.m.

What is the end time?

_____ : _____ p.m.

3. The start and end times are shown for a race. How long did the race take?

_____ minutes

Start (a.m.)

End (p.m.)

Solve Problems Involving Time

Name ____

Solve.

1. Ave went to dinner to at 6:15 p.m. She spent 22 minutes showering and dressing and spent 11 minutes drying and fixing her hair to get ready for dinner. What time did Ave begin getting ready for dinner? ____________

2. Adam went to school at 8:25 a.m. He spent 55 minutes bathing, shaving, and packing his backpack for school. He spent 23 minutes dressing and eating breakfast. What time did Adam begin getting ready for school? ____________

3. Lina ate lunch at 12:25 p.m. She spent 35 minutes cooking spaghetti and sauce and 18 cleaning up and setting the table. What time did Lina begin getting ready for lunch?

____

4. Hana ate a snack at 2:28 p.m. She spent 27 minutes making popcorn and pouring juice. She spent 12 minutes cleaning up the popcorn pot and washing up. What time did Hana begin getting ready eat a snack? ____________

5. Ajay ate lunch at 11:15 a.m. He spent 24 minutes grilling a sandwich and making fruit salad. He spent 15 minutes cleaning up his food prep area and setting the table. What time did Ajay begin getting ready for lunch? ____________

Understand Scaled Picture Graphs

Name ____________________

Review

You can make a scaled picture graph to show the number of hours each student practiced their musical instrument.

Hours Practiced	
Elaine	10
Connor	4
Elliot	6
Tristan	2

Hours of Practice of Rain

Elaine ♩ ♩ ♩ ♩ ♩
Connor ♩ ♩
Elliot ♩ ♩ ♩
Tristan ♩

Key: ♩ = 2 hours

Since each music note is 2 hours, the graph shows music notes to represent the number of hours practiced by each person in the table.

HINT
The key shows the value of each picture, or the scale.

The scaled picture graph shows the number of student votes for pets. Use the graph to answer the questions.

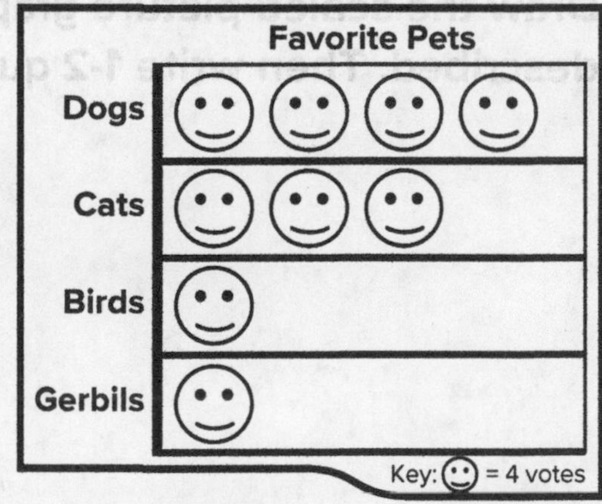

1. How many votes are represented by each picture? How do you know?

2. How many students voted for dogs? How do you know?

3. How many students voted for dogs, cats, birds, and gerbils? How do you know?

Understand Scaled Picture Graphs

Name ______________________________

Describe 3 different ways you could design the scale in a scaled picture graph to represent the data shown.

Votes for Favorite Fruit	
Apple	12
Pear	8
Mango	24
Orange	48
Kumquat	36

Draw the scaled picture graph using one of the ways you described. Then write 1-2 questions about the graph.

Understand Scaled Bar Graphs

Name ______________________________

Review

You can make a scaled bar graph to represent large amounts of data. The bars represent each number in the table. Each interval represents 3 votes.

HINT
The scale is greater than 1 on a scaled bar graph.

Favorite Sports	
Sport	**Number of Votes**
Baseball	6
Golf	3
Soccer	12
Football	12

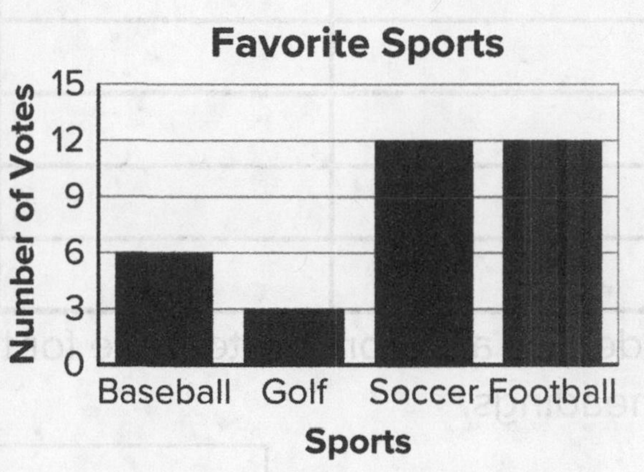

Complete the bar graph using the data in the table.

Favorite Yogurt Flavors	
Flavor	**Number of Students**
Strawberry	15
Lemon	20
Vanilla	30
Blueberry	15

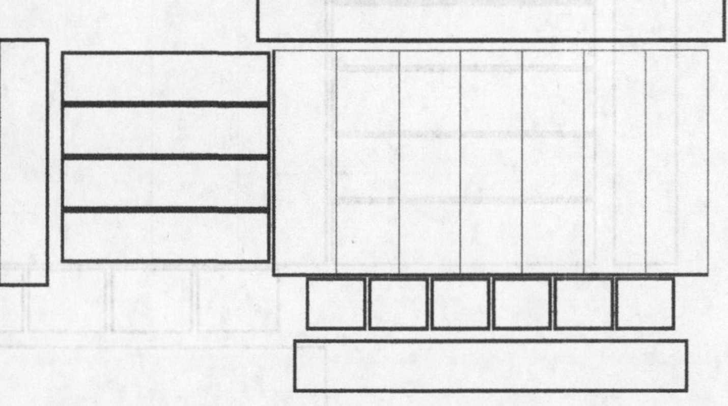

1. How did you decide the scale of your graph? Explain your reasoning.

Understand Scaled Bar Graphs

Name ______________________________

Complete the table with 4 categories that may be used to create a scaled bar graph.

Identify an appropriate scale for the bar graph. Include titles and headings.

Write and answer one question about the data in the graph.

Solve Problems Involving Scaled Graphs

Name ______________________________

Review

To solve problems involving scaled graphs, determine the scale. Then determine the numbers represented by each bar or symbol.

HINT
Write an equation for each part of the problem.

Example How many more tuna sandwiches did students eat than cheese and turkey?

500 + 100 = 600

650 − 600 = 50

Students ate 50 more tuna sandwiches than cheese and turkey.

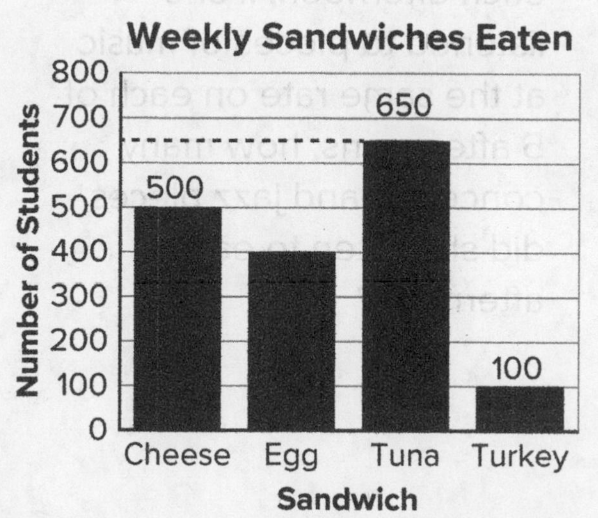

Use the bar graph to answer the question.

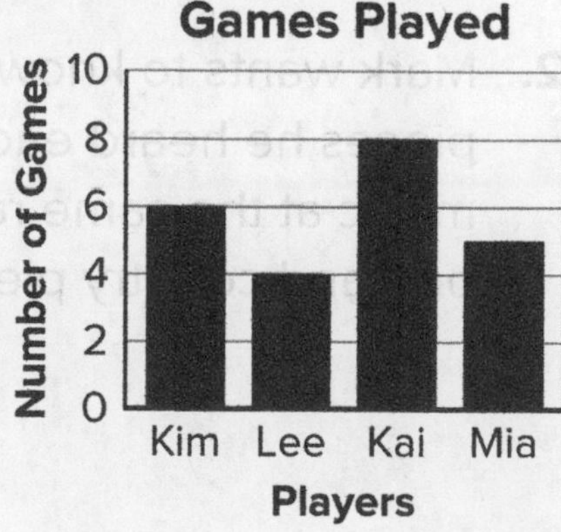

1. How many more games did Kai play than Kim? ____________

2. What is the difference between the greatest number of games played and the least number of games played? ____________

Solve Problems Involving Scaled Graphs

Name ____

Use the bar graph. Solve. Explain the steps you used to solve the problem.

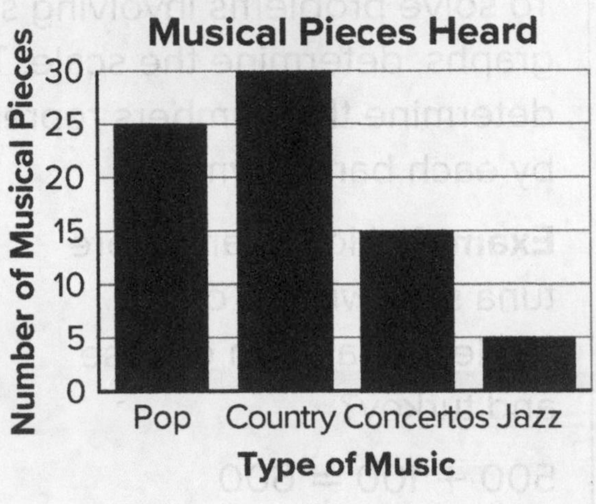

1. Maya wants to know how many concerto and jazz musical pieces she heard each afternoon. If she listened to pieces of music at the same rate on each of 5 afternoons, how many concertos and jazz pieces did she listen to each afternoon?

2. Mark wants to know how many pop and country musical pieces he heard each morning. If he listened to pieces of music at the same rate on each of 5 morning, how many pop and country pieces did he listen to each morning?

Lesson **12-10 • Reinforce Understanding**

Measure to Halves or Fourths of an Inch

Name ______________________________

Review

You can use a ruler to measure the length of objects to the nearest half inch or quarter inch.

HINT
Place the object at the 0 mark, and then find the nearest half inch or quarter inch at the other end.

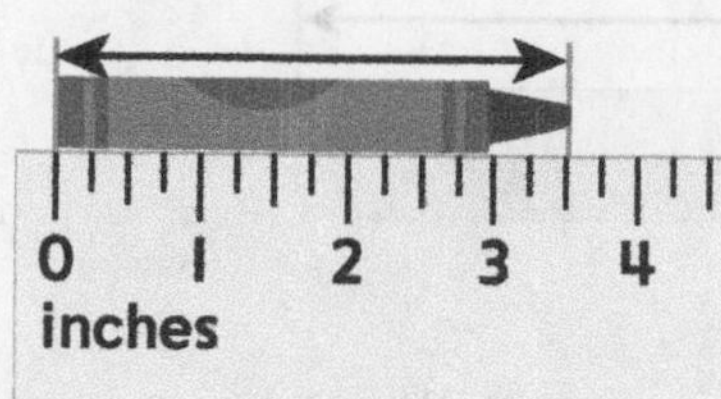

$3\frac{1}{2}$ inches

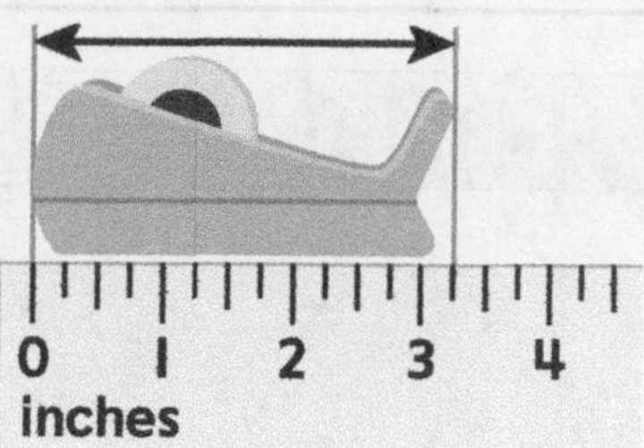

$3\frac{1}{4}$ inches

Use an inch ruler to measure each item to the nearest half inch.

1.

____

2.

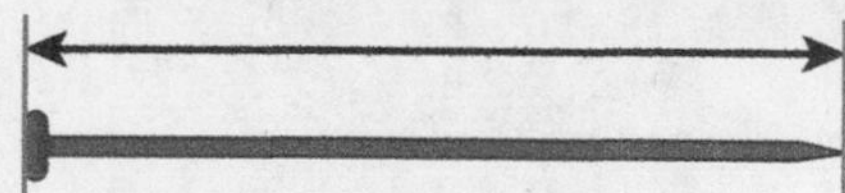

____

Use an inch ruler to measure each item to the nearest quarter inch.

3.

____

4.

____

Measure to Halves or Fourths of an Inch

Name ________________________________

Use an inch ruler to measure the pairs of objects.

Find the difference in their lengths. Explain your answer.

1.

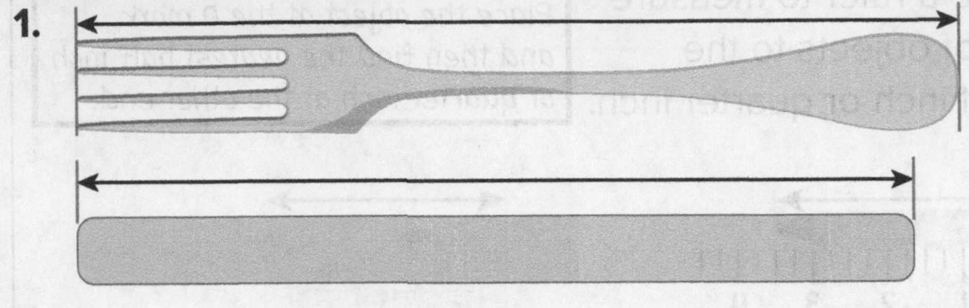

2.

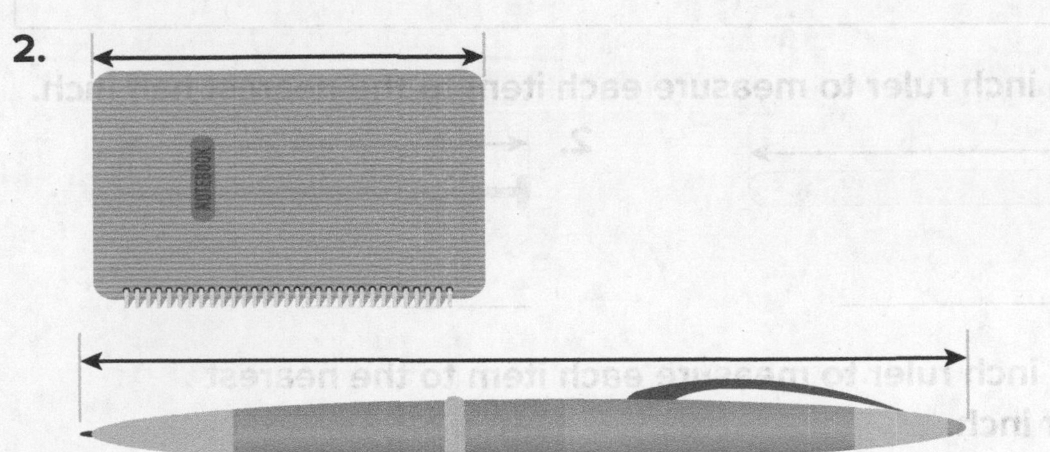

Lesson 12-11 • Reinforce Understanding

Show Measurement Data on a Line Plot

Name ________________________________

Review

You can create a line plot from measurement data. Each X can represent one data point. Create a line plot from the data in the tally table.

Lengths of Ribbon (in.)	
Length (in.)	Number
4	
$4\frac{1}{4}$	~~||||~~ |
$4\frac{1}{2}$	|||
$4\frac{3}{4}$	||
5	~~||||~~
$5\frac{1}{4}$	|
$5\frac{1}{2}$	||||
$5\frac{3}{4}$	~~||||~~
6	|

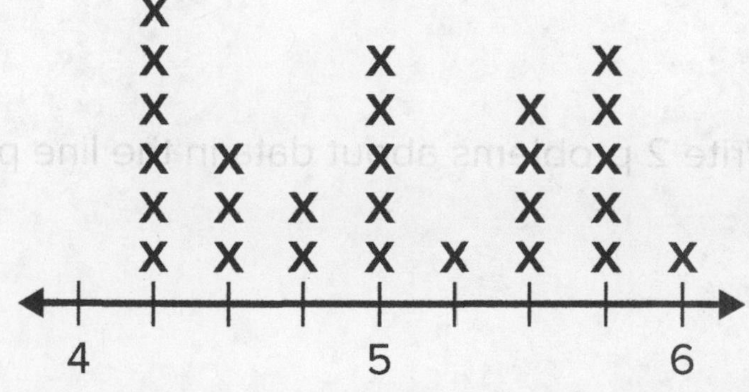

Use the line plot above to answer the questions.

1. How many ribbons are represented? _____

2. Which length is the most common? Which length is not represented in the collection? ______________________

3. How many ribbons are less than 5 inches but longer than 4 inches?

Show Measurement Data on a Line Plot

Name ____

The line plot shows the data in the tally table. Explain how to use the line plot to help someone understand the data.

Heights of Plants (in.)	
Height (in.)	**Number**
6	𝍷𝍷𝍷
8	𝍸 𝍷
10	𝍷𝍷
12	𝍷𝍷𝍷𝍷
18	𝍷

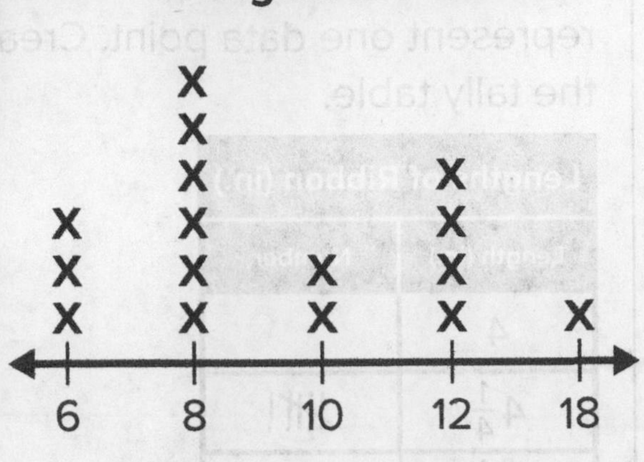

Write 2 problems about data in the line plot. Solve and explain.

Describe and Classify Polygons

Name ______________________________

Review

You can name and classify polygons based on their shared attributes.

HINT:
A polygon is a closed 2-dimensional figure formed by 3 or more straight sides that do not cross.

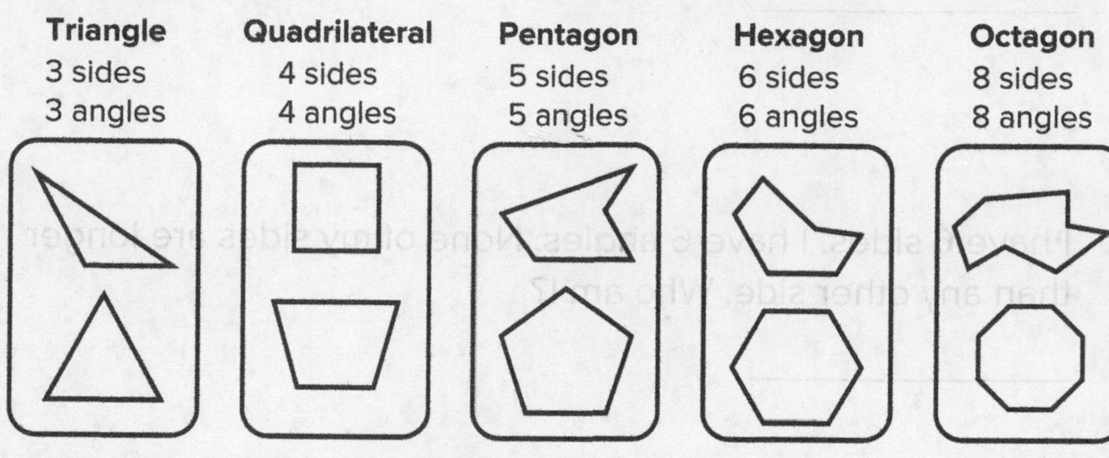

Write yes if the shape is a polygon. Write no if not, and tell why.

1. 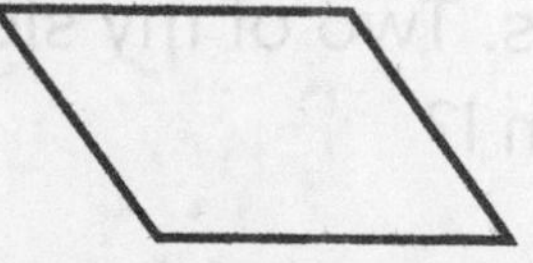

2. 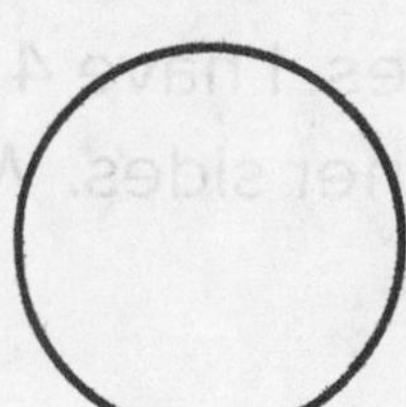

3. 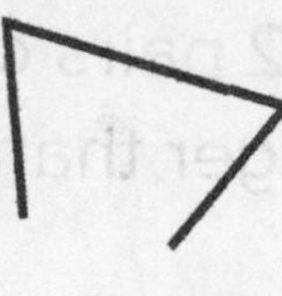

Classify the polygon.

4. ______________

5. ______________

6. ______________

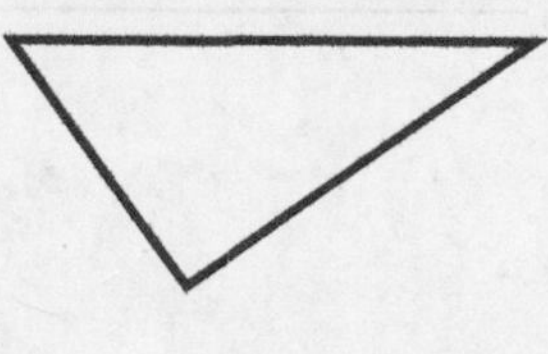

Describe and Classify Polygons

Name ______________________________

Solve the riddle. Write and draw the shape.

1. I have 4 sides. I have 4 angles. None of my sides are longer than any other side. Who am I?

____

2. I have 6 sides. I have 6 angles. None of my sides are longer than any other side. Who am I?

____

3. I have 2 pairs of equal sides. I have 4 angles. Two of my sides are longer than my two other sides. Who am I?

____

4. I have 8 sides and 8 angles. All of my sides are straight but some of my sides are longer than my other sides. Who am I?

____

Lesson **13-2 • Reinforce Understanding**

Describe Quadrilaterals

Name ______________________________

Review

You can describe quadrilaterals based on side lengths and angles.

Examples

4 equal sides

4 right angles

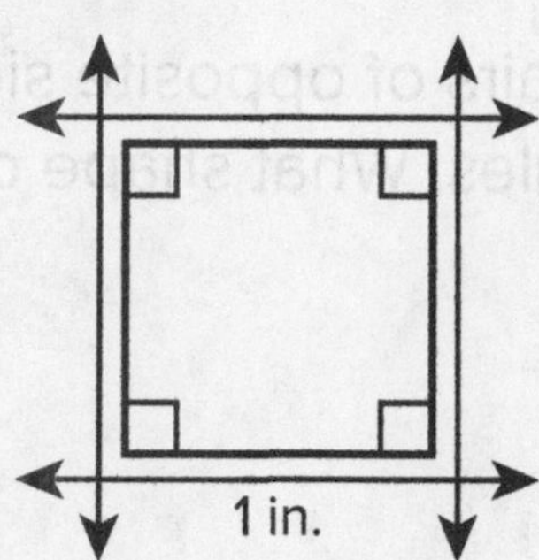

1 pair of opposite sides are equal

0 right angles

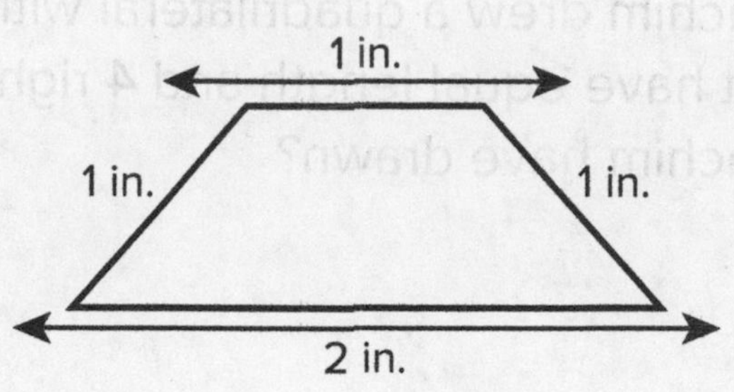

Describe the quadrilateral using the number of parallel sides, side lengths, and angles.

1.

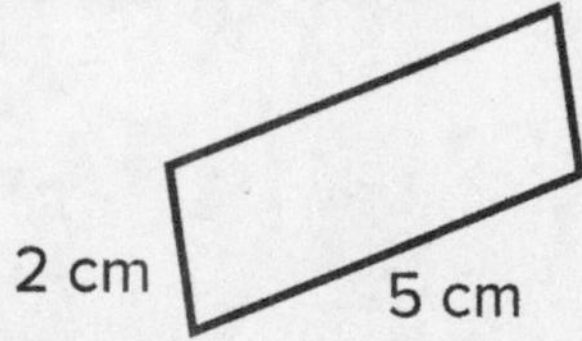

_______ pair(s) of equal sides

_______ right angle(s)

2.

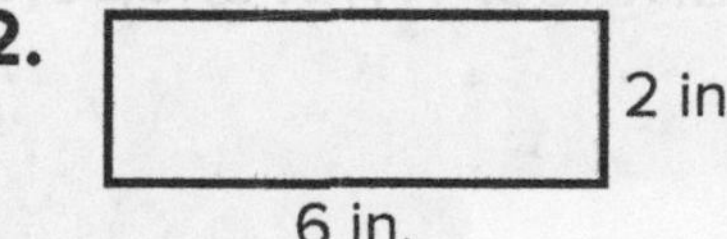

_______ pair(s) of equal sides

_______ right angle(s)

Circle the shape.

3. I am a quadrilateral with 1 pair of equal sides and 0 right angles.

Describe Quadrilaterals

Name ______________________________

Solve. Draw the quadrilateral.

1. Kaila drew a quadrilateral with 2 pairs of equal sides and 0 right angles. What shape could Kaila have drawn?

2. Joachim drew a quadrilateral with 2 pairs of opposite sides that have equal length and 4 right angles. What shape could Joachim have drawn?

3. Brooke drew a quadrilateral with 4 pairs of equal sides and 0 right angles. What shape could Brooke have drawn?

4. Mark drew a quadrilateral with 1 pair of equal sides and 0 right angles. What shape could Mark have drawn?

Classify Quadrilaterals

Name ________________________________

Review

You can classify quadrilaterals based on their shared attributes.

A quadrilateral with 4 right angles is a **rectangle.**

A quadrilateral with 4 equal sides is a **rhombus.**

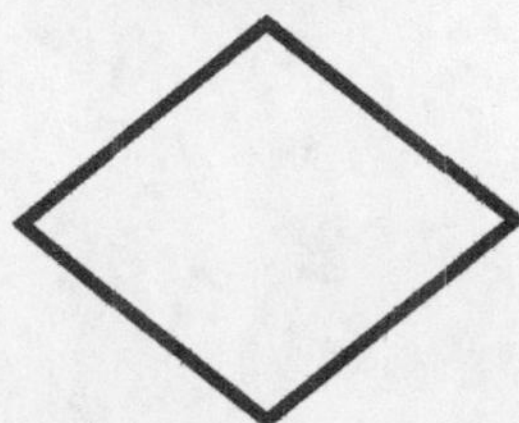

Write *square*, *rectangle*, *rhombus*, and *other* to label the quadrilaterals. Some have more than one label.

1. 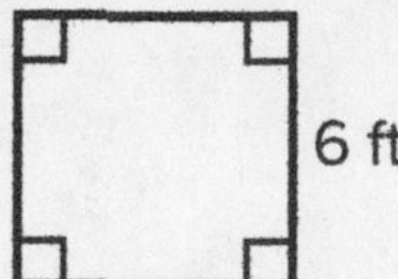

2. 

3. 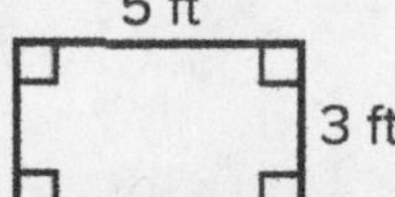

Classify Quadrilaterals

Name ______________________________

Draw 2 or more shapes for each category. Be sure to label the side lengths. Justify why each shape belongs in the category.

Rhombus

Rectangle

Other Quadrilateral

Draw Quadrilaterals with Specific Attributes

Name ____________________

Review

You can use side lengths and angles to draw quadrilaterals.

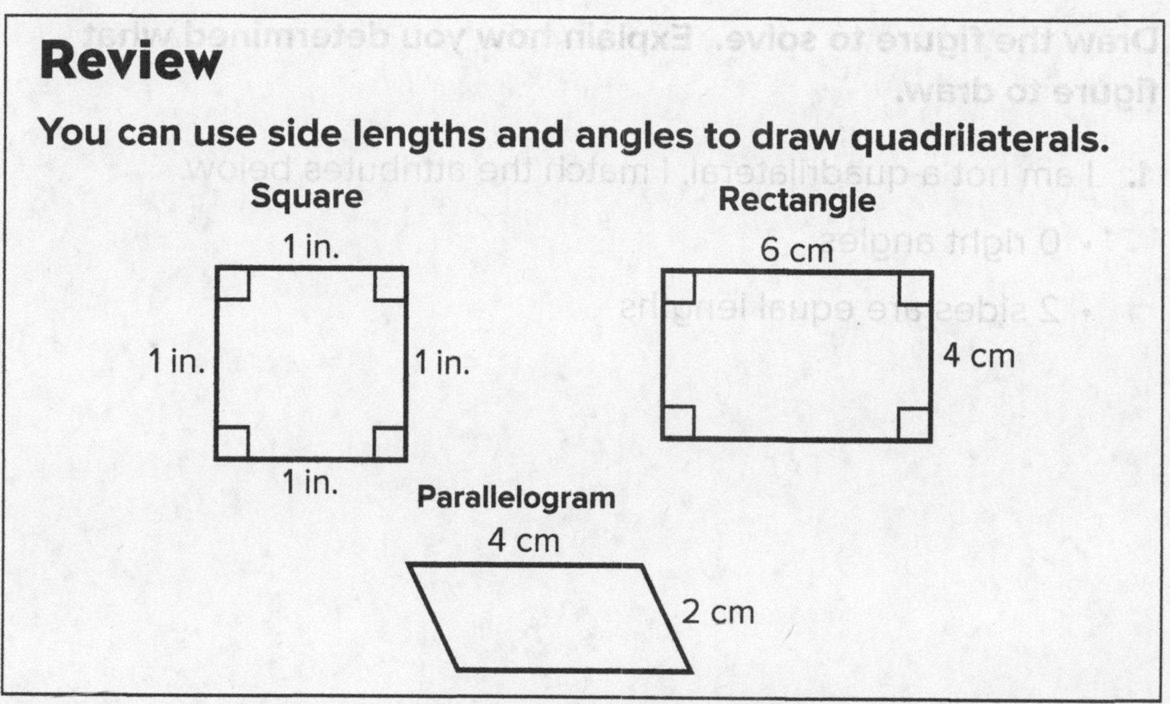

Draw a quadrilateral to match the description.

1. 2 pairs of equal sides and 4 right angles

2. 2 pairs of equal sides and 0 right angles

3. 4 equal sides and 0 right angles

4. 1 pair of opposite sides that are equal

Draw Quadrilaterals with Specific Attributes

Name ________________________________

Draw the figure to solve. Explain how you determined what figure to draw.

1. I am not a quadrilateral. I match the attributes below.

- 0 right angles
- 2 sides are equal lengths

2. I am a quadrilateral. I match the attributes below.

- 0 sides of equal length
- 0 right angles